STUDENT SOLUTIONS MANUAL

to accompany

CALCULUS

SINGLE VARIABLE   **THIRD EDITION**

Deborah Hughes-Hallett
University of Arizona

Andrew M. Gleason
Harvard University

William G. McCallum
University of Arizona

et al.

JOHN WILEY & SONS, INC.
New York • Chichester • Weinheim • Brisbane • Singapore • Toronto

COVER PHOTO   © Eddie Hironaka/The Image Bank.

To order books or for customer service call 1-800-CALL-WILEY (225-5945).

ISBN 0-471-44189-9

Printed in the United States of America

10 9 8 7 6 5 4 3 2

Printed and bound by Courier Kendallville, Inc.

CONTENTS

CHAPTER ONE

Solutions for Section 1.1

Exercises

1. $f(35)$ means the value of P corresponding to $t = 35$. Since t represents the number of years since 1950, we see that $f(35)$ means the population of the city in 1985. So, in 1985, the city's population was 12 million.

5. Rewriting the equation as
$$y = -\frac{12}{7}x + \frac{2}{7}$$
shows that the line has slope $-12/7$ and vertical intercept $2/7$.

9. The slope is $(3 - 2)/(2 - 0) = 1/2$. So the equation of the line is $y = (1/2)x + 2$.

13. The line parallel to $y = mx + c$ also has slope m, so its equation is
$$y = m(x - a) + b.$$
The line perpendicular to $y = mx + c$ has slope $-1/m$, so its equation will be
$$y = -\frac{1}{m}(x - a) + b.$$

17. Since the function goes from $x = -2$ to $x = 2$ and from $y = -2$ to $y = 2$, the domain is $-2 \leq x \leq 2$ and the range is $-2 \leq y \leq 2$.

21. The value of $f(t)$ is real provided $t^2 - 16 \geq 0$ or $t^2 \geq 16$. This occurs when either $t \geq 4$, or $t \leq -4$. Solving $f(t) = 3$, we have
$$\sqrt{t^2 - 16} = 3$$
$$t^2 - 16 = 9$$
$$t^2 = 25$$
so
$$t = \pm 5.$$

25. We know that N is proportional to $1/l^2$, so
$$N = \frac{k}{l^2}, \quad \text{for some constant } k.$$

Problems

29.

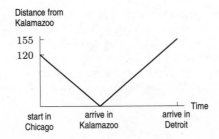

33. (a) This could be a linear function because w increases by 5 as h increases by 1.

(b) We find the slope m and the intercept b in the linear equation $w = b + mh$. We first find the slope m using the first two points in the table. Since we want w to be a function of h, we take

$$m = \frac{\Delta w}{\Delta h} = \frac{171 - 166}{69 - 68} = 5.$$

Substituting the first point and the slope $m = 5$ into the linear equation $w = b + mh$, we have $166 = b + (5)(68)$, so $b = -174$. The linear function is

$$w = 5h - 174.$$

The slope, $m = 5$ is in units of pounds per inch.

(c) We find the slope and intercept in the linear function $h = b + mw$ using $m = \Delta h / \Delta w$ to obtain the linear function

$$h = 0.2w + 34.8.$$

Alternatively, we could solve the linear equation found in part (b) for h. The slope, $m = 0.2$, has units inches per pound.

37. (a) $R = k(350 - H)$, where k is a positive constant.

If H is greater than $350°$, the rate is negative, indicating that a very hot yam will cool down toward the temperature of the oven.

(b) Letting H_0 equal the initial temperature of the yam, the graph of R against H looks like:

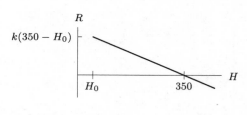

Note that by the temperature of the yam, we mean the average temperature of the yam, since the yam's surface will be hotter than its center.

Solutions for Section 1.2

Exercises

1. The graph shows a concave up function.

5. Initial quantity = 5; growth rate = $0.07 = 7\%$.

9. (a) The function is linear with initial population of 1000 and slope of 50, so $P = 1000 + 50t$.

(b) This function is exponential with initial population of 1000 and growth rate of 5%, so $P = 1000(1.05)^t$.

Problems

13. (a) Advertising is generally cheaper in bulk; spending more money will give better and better marginal results initially, (Spending $5,000 could give you a big newspaper ad reaching 200,000 people; spending $100,000 could give you a series of TV spots reaching 50,000,000 people.) A graph is shown below, left.

(b) The temperature of a hot object decreases at a rate proportional to the difference between its temperature and the temperature of the air around it. Thus, the temperature of a very hot object decreases more quickly than a cooler object. The graph is decreasing and concave up. (We are assuming that the coffee is all at the same temperature.)

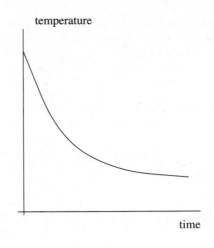

17. (a) Using $Q = Q_0(1 - r)^t$ for loss, we have

$$Q = 10{,}000(1 - 0.1)^{10} = 10{,}000(0.9)^{10} = 3486.78.$$

The investment was worth $3486.78 after 10 years.

(b) Measuring time from the moment at which the stock begins to gain value and letting $Q_0 = 3486.78$, the value after t years is

$$Q = 3486.78(1 + 0.1)^t = 3486.78(1.1)^t.$$

We can estimate the value of t when $Q = 10{,}000$ by tracing along a graph of Q, giving $t \approx 11$. It will take about 11 years to get the investment back to $10,000.

21. The difference, D, between the horizontal asymptote and the graph appears to decrease exponentially, so we look for an equation of the form

$$D = D_0 a^x$$

where $D_0 = 4 = $ difference when $x = 0$. Since $D = 4 - y$, we have

$$4 - y = 4a^x \quad \text{or} \quad y = 4 - 4a^x = 4(1 - a^x)$$

The point $(1, 2)$ is on the graph, so $2 = 4(1 - a^1)$, giving $a = \frac{1}{2}$.
Therefore $y = 4(1 - (\frac{1}{2})^x) = 4(1 - 2^{-x})$.

25. Since $e^{-0.5t} = (e^{-0.5})^t \approx (0.61)^t$, we have $P = 2(0.61)^t$. This is exponential decay since -0.5 is negative. We can also see that this is decay because $0.61 < 1$.

29. Direct calculation reveals that each 1000 foot increase in altitude results in a longer takeoff roll by a factor of about 1.096. Since the value of d when $h = 0$ (sea level) is $d = 670$, we are led to the formula

$$d = 670(1.096)^{h/1000},$$

where d is the takeoff roll, in feet, and h is the airport's elevation, in feet.

Alternatively, we can write

$$d = d_0 a^h,$$

where d_0 is the sea level value of d, $d_0 = 670$. In addition, when $h = 1000$, $d = 734$, so

$$734 = 670a^{1000}.$$

Solving for a gives

$$a = \left(\frac{734}{670}\right)^{1/1000} = 1.00009124,$$

so

$$d = 670(1.00009124)^h.$$

33. (a) This is the graph of a linear function, which increases at a constant rate, and thus corresponds to $k(t)$, which increases by 0.3 over each interval of 1.

(b) This graph is concave down, so it corresponds to a function whose increases are getting smaller, as is the case with $h(t)$, whose increases are 10, 9, 8, 7, and 6.

(c) This graph is concave up, so it corresponds to a function whose increases are getting bigger, as is the case with $g(t)$, whose increases are 1, 2, 3, 4, and 5.

37. Because the population is growing exponentially, the time it takes to double is the same, regardless of the population levels we are considering. For example, the population is 20,000 at time 3.7, and 40,000 at time 6.0. This represents a doubling of the population in a span of $6.0 - 3.7 = 2.3$ years.

How long does it take the population to double a second time, from 40,000 to 80,000? Looking at the graph once again, we see that the population reaches 80,000 at time $t = 8.3$. This second doubling has taken $8.3 - 6.0 = 2.3$ years, the same amount of time as the first doubling.

Further comparison of any two populations on this graph that differ by a factor of two will show that the time that separates them is 2.3 years. Similarly, during any 2.3 year period, the population will double. Thus, the doubling time is 2.3 years.

Suppose $P = P_0 a^t$ doubles from time t to time $t + d$. We now have $P_0 a^{t+d} = 2P_0 a^t$, so $P_0 a^t a^d = 2P_0 a^t$. Thus, canceling P_0 and a^t, d must be the number such that $a^d = 2$, no matter what t is.

Solutions for Section 1.3

Exercises

1. **(a)** $g(2 + h) = (2 + h)^2 + 2(2 + h) + 3 = 4 + 4h + h^2 + 4 + 2h + 3 = h^2 + 6h + 11$.
   (b) $g(2) = 2^2 + 2(2) + 3 = 4 + 4 + 3 = 11$, which agrees with what we get by substituting $h = 0$ into (a).
   (c) $g(2 + h) - g(2) = (h^2 + 6h + 11) - (11) = h^2 + 6h$.

5. $m(z + 1) - m(z) = (z + 1)^2 - z^2 = 2z + 1$.

9. **(a)** $f(25)$ is q corresponding to $p = 25$, or, in other words, the number of items sold when the price is 25.
   (b) $f^{-1}(30)$ is p corresponding to $q = 30$, or the price at which 30 units will be sold.

13. The function is not invertible since there are many horizontal lines which hit the function twice.

17. This looks like a shift of the graph $y = x^3$. The graph is shifted to the right 2 units and down 1 unit, so a possible formula is $y = (x - 2)^3 - 1$.

Problems

21. Not invertible. Given a certain number of customers, say $f(t) = 1500$, there could be many times, t, during the day at which that many people were in the store. So we don't know which time instant is the right one.

25. $f(g(1)) = f(2) \approx 0.4$.

29. Using the same way to compute $g(f(x))$ as in Problem 26, we get the following table. Then we can plot the graph of $g(f(x))$.

x	$f(x)$	$g(f(x))$
-3	3	-2.6
-2.5	0.1	0.8
-2	-1	-1.4
-1.5	-1.3	-1.8
-1	-1.2	-1.7
-0.5	-1	-1.4
0	-0.8	-1
0.5	-0.6	-0.6
1	-0.4	-0.3
1.5	-0.1	0.3
2	0.3	1.1
2.5	0.9	2
3	1.6	2.2

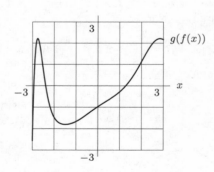

33. Since $B = y - 1$ and $n = 2B^2 - B$, substitution gives

$$n = 2B^2 - B = 2(y - 1)^2 - (y - 1) = 2y^2 - 5y + 3.$$

Solutions for Section 1.4

Exercises

1. The function e^x has a vertical intercept of 1, so must be A. The function $\ln x$ has an x-intercept of 1, so must be D. The graphs of x^2 and $x^{1/2}$ go through the origin. The graph of $x^{1/2}$ is concave down so it corresponds to graph C and the graph of x^2 is concave up so it corresponds to graph B.

5. Isolating the exponential term

$$20 = 50(1.04)^x$$
$$\frac{20}{50} = (1.04)^x$$

Taking logs of both sides

$$\log \frac{2}{5} = \log(1.04)^x$$
$$\log \frac{2}{5} = x \log(1.04)$$
$$x = \frac{\log(2/5)}{\log(1.04)} \approx -23.4.$$

9. $\ln(2^x) = \ln(e^{x+1})$
   $x \ln 2 = (x+1) \ln e$
   $x \ln 2 = x + 1$
   $0.693x = x + 1$
   $x = \dfrac{1}{0.693 - 1} \approx -3.26$

13. Using the rules for ln, we get

$$\ln 7^{x+2} = \ln e^{17x}$$
$$(x+2) \ln 7 = 17x$$
$$x(\ln 7 - 17) = -2 \ln 7$$
$$x = \frac{-2 \ln 7}{\ln 7 - 17} \approx 0.26.$$

17. $t = \dfrac{\log a}{\log b}$.

21. $t = \ln \dfrac{a}{b}$.

25. Using the identity $e^{\ln x} = x$, we have $5A^2$.

29. Since $(1.5)^t = \left(e^{\ln 1.5}\right)^t = e^{(\ln 1.5)t} = e^{0.41t}$, we have $P = 15e^{0.41t}$. Since 0.41 is positive, this is exponential growth.

33. If $p(t) = (1.04)^t$, then, for p^{-1} the inverse of p, we should have

$$(1.04)^{p^{-1}(t)} = t,$$
$$p^{-1}(t) \log(1.04) = \log t,$$
$$p^{-1}(t) = \frac{\log t}{\log(1.04)} \approx 58.708 \log t.$$

Problems

37. (a) The initial dose is 10 mg.
    (b) Since $0.82 = 1 - 0.18$, the decay rate is 0.18, so 18% leaves the body each hour.
    (c) When $t = 6$, we have $A = 10(0.82)^6 = 3.04$. The amount in the body after 6 hours is 3.04 mg.

(d) We want to find the value of t when $A = 1$. Using logarithms:

$$1 = 10(0.82)^t$$
$$0.1 = (0.82)^t$$
$$\ln(0.1) = t \ln(0.82)$$
$$t = 11.60 \text{ hours.}$$

After 11.60 hours, the amount is 1 mg.

41. In ten years, the substance has decayed to 40% of its original mass. In another ten years, it will decay by an additional factor of 40%, so the amount remaining after 20 years will be $100 \cdot 40\% \cdot 40\% = 16$ kg.

45. Let $t =$ number of years since 1980. Then the number of vehicles, V, in millions, at time t is given by

$$V = 170(1.04)^t$$

and the number of people, P, in millions, at time t is given by

$$P = 227(1.01)^t.$$

There is an average of one vehicle per person when $\dfrac{V}{P} = 1$, or $V = P$. Thus, we must solve for t the equation:

$$170(1.04)^t = 227(1.01)^t,$$

which implies

$$\left(\frac{1.04}{1.01}\right)^t = \frac{(1.04)^t}{(1.01)^t} = \frac{227}{170}$$

Taking logs on both sides,

$$t \log \frac{1.04}{1.01} = \log \frac{227}{170}.$$

Therefore,

$$t = \frac{\log\left(\frac{227}{170}\right)}{\log\left(\frac{1.04}{1.01}\right)} \approx 9.9 \text{ years.}$$

So there was, according to this model, about one vehicle per person in 1990.

49. Since the amount of strontium-90 remaining halves every 29 years, we can solve for the decay constant;

$$0.5P_0 = P_0 e^{-29k}$$
$$k = \frac{\ln(1/2)}{-29}.$$

Knowing this, we can look for the time t in which $P = 0.10P_0$, or

$$0.10P_0 = P_0 e^{\ln(0.5)t/29}$$
$$t = \frac{29 \ln(0.10)}{\ln(0.5)} = 96.34 \text{ years.}$$

Solutions for Section 1.5

Exercises

1.

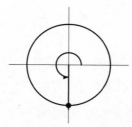

$$\sin\left(\frac{3\pi}{2}\right) = -1 \quad \text{is negative.}$$

$$\cos\left(\frac{3\pi}{2}\right) = 0$$

$$\tan\left(\frac{3\pi}{2}\right) \quad \text{is undefined.}$$

5.

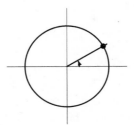

$$\sin\left(\frac{\pi}{6}\right) \quad \text{is positive.}$$

$$\cos\left(\frac{\pi}{6}\right) \quad \text{is positive.}$$

$$\tan\left(\frac{\pi}{6}\right) \quad \text{is positive.}$$

9. $-1 \text{ radian} \cdot \frac{180°}{\pi \text{ radians}} = -\left(\frac{180°}{\pi}\right) \approx -60°$

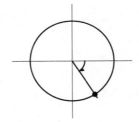

$$\sin(-1) \quad \text{is negative}$$

$$\cos(-1) \quad \text{is positive}$$

$$\tan(-1) \quad \text{is negative}$$

13. (a) We determine the amplitude of y by looking at the coefficient of the cosine term. Here, the coefficient is 1, so the amplitude of y is 1. Note that the constant term does not affect the amplitude.

(b) We know that the cosine function $\cos x$ repeats itself at $x = 2\pi$, so the function $\cos(3x)$ must repeat itself when $3x = 2\pi$, or at $x = 2\pi/3$. So the period of y is $2\pi/3$. Here as well the constant term has no effect.

(c) The graph of y is shown in the figure below.

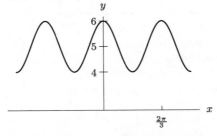

17. The graph is an inverted sine curve with amplitude 1 and period 2π, shifted up by 2, so it is given by $f(x) = 2 - \sin x$.

21. The graph is a sine curve which has been shifted up by 2, so $f(x) = (\sin x) + 2$.

25. The graph is a cosine curve with period $2\pi/5$ and amplitude 2, so it is given by $f(x) = 2\cos(5x)$.

Problems

29. Using the fact that 1 revolution $= 2\pi$ radians and 1 minute $= 60$ seconds, we have

$$200\frac{\text{rev}}{\text{min}} = (200) \cdot 2\pi\frac{\text{rad}}{\text{min}} = 200 \cdot 2\pi\frac{1}{60}\frac{\text{rad}}{\text{sec}}$$

$$\approx \frac{(200)(6.283)}{60}$$

$$\approx 20.94 \text{ radians per second.}$$

Similarly, 500 rpm is equivalent to 52.36 radians per second.

33. (a) Reading the graph of θ against t shows that $\theta \approx 5.2$ when $t = 1.5$. Since the coordinates of P are $x = 5\cos\theta$, $y = 5\sin\theta$, when $t = 1.5$ the coordinates are

$$(x, y) \approx (5\cos 5.2, 5\sin 5.2) = (2.3, -4.4).$$

(b) As t increases from 0 to 5, the angle θ increases from 0 to about 6.3 and then decreases to 0 again. Since $6.3 \approx 2\pi$, this means that P starts on the x-axis at the point $(5, 0)$, moves counterclockwise the whole way around the circle (at which time $\theta \approx 2\pi$), and then moves back clockwise to its starting point.

37. The US voltage has a maximum value of 156 volts and has a period of $1/60$ of a second, so it executes 60 cycles a second. The European voltage has a higher maximum of 339 volts, and a slightly longer period of $1/50$ seconds, so it oscillates at 50 cycles per second.

41.

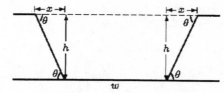

Figure 1.1

Figure 1.1 shows that the cross-sectional area is one rectangle of area hw and two triangles. Each triangle has height h and base x, where

$$\frac{h}{x} = \tan\theta \quad \text{so} \quad x = \frac{h}{\tan\theta}.$$

$$\text{Area of triangle} = \frac{1}{2}xh = \frac{h^2}{2\tan\theta}$$

$$\text{Total area} = \text{Area of rectangle} + 2(\text{Area of triangle})$$

$$= hw + 2 \cdot \frac{h^2}{2\tan\theta} = hw + \frac{h^2}{\tan\theta}.$$

Solutions for Section 1.6

Exercises

1. Exponential growth dominates power growth as $x \to \infty$, so $10 \cdot 2^x$ is larger.

5. (I) (a) Minimum degree is 3 because graph turns around twice.
   (b) Leading coefficient is negative because $y \to -\infty$ as $x \to \infty$.
(II) (a) Minimum degree is 4 because graph turns around three times.
   (b) Leading coefficient is positive because $y \to \infty$ as $x \to \infty$.
(III) (a) Minimum degree is 4 because graph turns around three times.
   (b) Leading coefficient is negative because $y \to -\infty$ as $x \to \infty$.
(IV) (a) Minimum degree is 5 because graph turns around four times.
   (b) Leading coefficient is negative because $y \to -\infty$ as $x \to \infty$.
(V) (a) Minimum degree is 5 because graph turns around four times.
   (b) Leading coefficient is positive because $y \to \infty$ as $x \to \infty$.

9. (a) A polynomial has the same end behavior as its leading term, so this polynomial behaves as $-5x^4$ globally. Thus we have:
$$f(x) \to -\infty \text{ as } x \to -\infty, \quad \text{and} \quad f(x) \to -\infty \text{ as } x \to +\infty.$$

(b) Polynomials behave globally as their leading term, so this rational function behaves globally as $(3x^2)/(2x^2)$, or $3/2$. Thus we have:
$$f(x) \to 3/2 \text{ as } x \to -\infty, \quad \text{and} \quad f(x) \to 3/2 \text{ as } x \to +\infty.$$

(c) We see from a graph of $y = e^x$ that
$$f(x) \to 0 \text{ as } x \to -\infty, \quad \text{and} \quad f(x) \to +\infty \text{ as } x \to +\infty.$$

Problems

13. $f(x) = k(x+2)(x-2)^2(x-5) = k(x^4 - 7x^3 + 6x^2 + 28x - 40)$, where $k < 0$. ($k \approx -\frac{1}{15}$ if the scales are equal; otherwise one can't tell how large k is.)

17. (a)   (i) The water that has flowed out of the pipe in 1 second is a cylinder of radius r and length 3 cm. Its volume is
$$V = \pi r^2(3) = 3\pi r^2.$$

(ii) If the rate of flow is k cm/sec instead of 3 cm/sec, the volume is given by
$$V = \pi r^2(k) = \pi r^2 k.$$

(b)   (i) The graph of V as a function of r is a quadratic. See Figure 1.2.

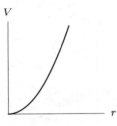

Figure 1.2

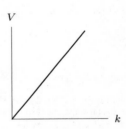

Figure 1.3

(ii) The graph of V as a function of k is a line. See Figure 1.3.

21. Let $D(v)$ be the stopping distance required by an Alpha Romeo as a function of its velocity. The assumption that stopping distance is proportional to the square of velocity is equivalent to the equation
$$D(v) = kv^2$$

where k is a constant of proportionality. To determine the value of k, we use the fact that $D(70) = 177$.
$$D(70) = k(70)^2 = 177.$$

Thus,
$$k = \frac{177}{70^2} \approx 0.0361.$$

It follows that
$$D(35) = \left(\frac{177}{70^2}\right)(35)^2 = \frac{177}{4} = 44.25 \text{ ft}$$

and
$$D(140) = \left(\frac{177}{70^2}\right)(140)^2 = 708 \text{ ft.}$$

Thus, at half the speed it requires one fourth the distance, whereas at twice the speed it requires four times the distance, as we would expect from the equation. (We could in fact have figured it out that way, without solving for k explicitly.)

25. (a)   (i) If $(1, 1)$ is on the graph, we know that
$$1 = a(1)^2 + b(1) + c = a + b + c.$$

(ii) If $(1, 1)$ is the vertex, then the axis of symmetry is $x = 1$, so

$$-\frac{b}{2a} = 1,$$

and thus

$$a = -\frac{b}{2}, \text{ so } b = -2a.$$

But to be the vertex, $(1, 1)$ must also be on the graph, so we know that $a + b + c = 1$. Substituting $b = -2a$, we get $-a + c = 1$, which we can rewrite as $a = c - 1$, or $c = 1 + a$.

(iii) For $(0, 6)$ to be on the graph, we must have $f(0) = 6$. But $f(0) = a(0^2) + b(0) + c = c$, so $c = 6$.

(b) To satisfy all the conditions, we must first, from (a)(iii), have $c = 6$. From (a)(ii), $a = c - 1$ so $a = 5$. Also from (a)(ii), $b = -2a$, so $b = -10$. Thus the completed equation is

$$y = f(x) = 5x^2 - 10x + 6,$$

which satisfies all the given conditions.

29. Consider the end behavior of the graph; that is, as $x \to +\infty$ and $x \to -\infty$. The ends of a degree 5 polynomial are in Quadrants I and III if the leading coefficient is positive or in Quadrants II and IV if the leading coefficient is negative. Thus, there must be at least one root. Since the degree is 5, there can be no more than 5 roots. Thus, there may be 1, 2, 3, 4, or 5 roots. Graphs showing these five possibilities are shown in Figure 1.4.

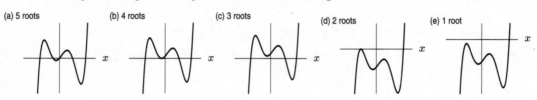

Figure 1.4

33. The graphs of both these functions will resemble that of x^3 on a large enough window. One way to tackle the problem is to graph them both (along with x^3 if you like) in successively larger windows until the graphs come together. In Figure 1.5, f, g and x^3 are graphed in four windows. In the largest of the four windows the graphs are indistinguishable, as required. Answers may vary.

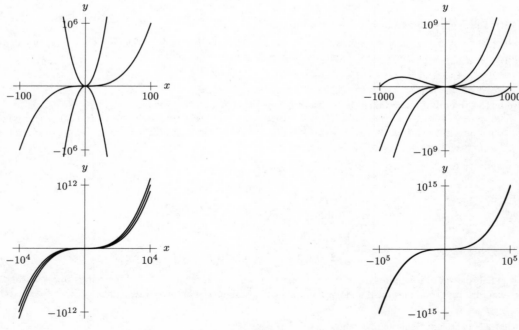

Figure 1.5

Solutions for Section 1.7

Exercises

1. Yes, because $2x + x^{2/3}$ is defined for all x.

5. Yes, because $2x - 5$ is positive for $3 \le x \le 4$.

9. No, because $e^x - 1 = 0$ at $x = 0$.

Problems

13. For any value of k, the function is continuous at every point except $x = 2$. We choose k to make the function continuous at $x = 2$.

Since $3x^2$ takes the value $3(2^2) = 12$ at $x = 2$, we choose k so that kx goes through the point $(2, 12)$. Thus $k = 6$.

17.

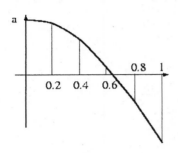 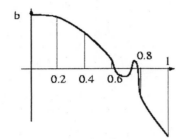 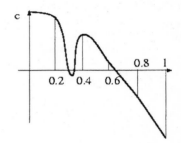

Solutions for Chapter 1 Review

Exercises

1. (a) The domain of f is the set of values of x for which the function is defined. Since the function is defined by the graph and the graph goes from $x = 0$ to $x = 7$, the domain of f is $[0, 7]$.

   (b) The range of f is the set of values of y attainable over the domain. Looking at the graph, we can see that y gets as high as 5 and as low as -2, so the range is $[-2, 5]$.

   (c) Only at $x = 5$ does $f(x) = 0$. So 5 is the only root of $f(x)$.

   (d) Looking at the graph, we can see that $f(x)$ is decreasing on $(1, 7)$.

   (e) The graph indicates that $f(x)$ is concave up at $x = 6$.

   (f) The value $f(4)$ is the y-value that corresponds to $x = 4$. From the graph, we can see that $f(4)$ is approximately 1.

   (g) This function is not invertible, since it fails the horizontal-line test. A horizontal line at $y = 3$ would cut the graph of $f(x)$ in two places, instead of the required one.

5. The amplitude is 2. The period is $2\pi/5$. See Figure 1.6.

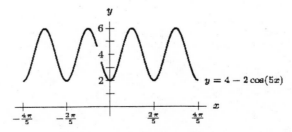

Figure 1.6

9. $y = -kx(x + 5) = -k(x^2 + 5x)$, where $k > 0$ is any constant.

13. $x = ky(y - 4) = k(y^2 - 4y)$, where $k > 0$ is any constant.

17. There are many solutions for a graph like this one. The simplest is $y = 1 - e^{-x}$, which gives the graph of $y = e^x$, flipped over the x-axis and moved up by 1. The resulting graph passes through the origin and approaches $y = 1$ as an upper bound, the two features of the given graph.

21. The graph appears to have a vertical asymptote at $t = 0$, so $f(t)$ is not continuous on $[-1, 1]$.

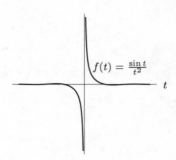

Problems

25. We will let

$$
\begin{aligned}
T &= \text{amount of fuel for take-off,} \\
L &= \text{amount of fuel for landing,} \\
P &= \text{amount of fuel per mile in the air,} \\
m &= \text{the length of the trip in miles.}
\end{aligned}
$$

Then Q, the total amount of fuel needed, is given by

$$Q(m) = T + L + Pm.$$

29. Assuming the US population grows exponentially, we have

$$248.7 = 226.5e^{10k}$$
$$k = \frac{\ln(1.098)}{10} = 0.00935.$$

We want to find the time t in which

$$300 = 226.5e^{0.00935t}$$
$$t = \frac{\ln(1.324)}{0.00935} = 30 \text{ years.}$$

Thus, the population will go over 300 million around the year 2010.

33. (a) Let the height of the can be h. Then

$$V = \pi r^2 h.$$

The surface area consists of the area of the ends (each is πr^2) and the curved sides (area $2\pi rh$), so

$$S = 2\pi r^2 + 2\pi rh.$$

Solving for h from the formula for V, we have

$$h = \frac{V}{\pi r^2}.$$

Substituting into the formula for S, we get

$$S = 2\pi r^2 + 2\pi r \cdot \frac{V}{\pi r^2} = 2\pi r^2 + \frac{2V}{r}.$$

(b) For large r, the $2V/r$ term becomes negligible, meaning $S \approx 2\pi r^2$, and thus $S \to \infty$ as $r \to \infty$.

(c) The graph is in Figure 1.7.

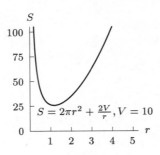

Figure 1.7

37. **(a)** Yes, f is invertible, since f is increasing everywhere.

(b) $f^{-1}(400)$ is the year in which 400 million motor vehicles were registered in the world. From the picture, we see that $f^{-1}(400)$ is around 1979.

(c) Since the graph of f^{-1} is the reflection of the graph of f over the line $y = x$, we get Figure 1.8.

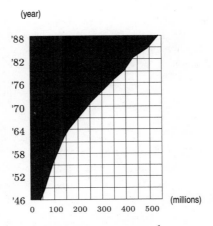

Figure 1.8: Graph of f^{-1}

41. **(a)** $r(p) = kp(A - p)$, where $k > 0$ is a constant.

(b) $p = A/2$.

CAS Challenge Problems

45. **(a)** As $x \to \infty$, the term e^{6x} dominates and tends to ∞. Thus, $f(x) \to \infty$ as $x \to \infty$.

As $x \to -\infty$, the terms of the form e^{kx}, where $k = 6, 5, 4, 3, 2, 1$, all tend to zero. Thus, $f(x) \to 16$ as $x \to -\infty$.

(b) A CAS gives

$$f(x) = (e^x + 1)(e^{2x} - 2)(e^x - 2)(e^{2x} + 2e^x + 4).$$

Since e^x is always positive, the factors $(e^x + 1)$ and $(e^{2x} + 2e^x + 4)$ are never zero. The other factors each lead to a zero, so there are two zeros.

(c) The zeros are given by

$$e^{2x} = 2 \quad \text{so} \quad x = \frac{\ln 2}{2}$$

$$e^x = 2 \quad \text{so} \quad x = \ln 2.$$

Thus, one zero is twice the size of the other.

49. Using the trigonometric expansion capabilities of your computer algebra system, you get something like

$$\cos(4x) = \cos^4(x) - 6\cos^2(x)\sin^2(x) + \sin^4(x).$$

Answers may vary.

(a) To get rid of the powers of cosine, use the identity $\cos^2(x) = 1 - \sin^2(x)$. This gives

$$\cos(4x) = \cos^4(x) - 6\cos^2(x)\left(1 - \cos^2(x)\right) + \left(1 - \cos^2(x)\right)^2.$$

Finally, using the CAS to simplify,

$$\cos(4x) = 1 - 8\cos^2(x) + 8\cos^4(x).$$

(b) This time we use $\sin^2(x) = 1 - \cos^2(x)$ to get rid of powers of sine. We get

$$\cos(4x) = \left(1 - \sin^2(x)\right)^2 - 6\sin^2(x)\left(1 - \sin^2(x)\right) + \sin^4(x) = 1 - 8\sin^2(x) + 8\sin^4(x).$$

CHECK YOUR UNDERSTANDING

1. False. A line can be put through any two points in the plane. However, if the line is vertical, it is not the graph of a function.

5. True. The highest degree term in a polynomial determines how the polynomial behaves when x is very large in the positive or negative direction. When n is odd, x^n is positive when x is large and positive but negative when x is large and negative. Thus if a polynomial $p(x)$ has odd degree, it will be positive for some values of x and negative for other values of x. Since every polynomial is continuous, the Intermediate Value Theorem then guarantees that $p(x) = 0$ for some value of x.

9. False. Suppose $y = 5^x$. Then increasing x by 1 increases y by a factor of 5. However increasing x by 2 increases y by a factor of 25, not 10, since

$$y = 5^{x+2} = 5^x \cdot 5^2 = 25 \cdot 5^x.$$

(Other examples are possible.)

13. True. The period is $2\pi/(200\pi) = 1/100$ seconds. Thus, the function executes 100 cycles in 1 second.

17. False. For $x < 0$, as x increases, x^2 decreases, so e^{-x^2} increases.

21. True. If $b > 1$, then $ab^x \to 0$ as $x \to -\infty$. If $0 < b < 1$, then $ab^x \to 0$ as $x \to \infty$. In either case, the function $y = a + ab^x$ has $y = a$ as the horizontal asymptote.

25. False. A counterexample is given by $f(x) = x^2$ and $g(x) = x + 1$. The function $f(g(x)) = (x+1)^2$ is not even because $f(g(1)) = 4$ and $f(g(-1)) = 0 \neq 4$.

29. Let $f(x) = \dfrac{1}{(x-1)(x-2)(x-3)\cdots(x-16)(x-17)}$. This function has an asymptote corresponding to every factor in the denominator. Other answers are possible.

33. This is impossible. If $a < b$, then $f(a) < f(b)$, since f is increasing, and $g(a) > g(b)$, since g is decreasing, so $-g(a) < -g(b)$. Therefore, if $a < b$, then $f(a) - g(a) < f(b) - g(b)$, which means that $f(x) + g(x)$ is increasing.

37. False. For example, let $f(x) = \log x$. Then $f(x)$ is increasing on $[1, 2]$, but $f(x)$ is concave down. (Other examples are possible.)

41. False. For example, let $f(x) = \begin{cases} 1 & x \leq 3 \\ 2 & x > 3 \end{cases}$, then $f(x)$ is defined at $x = 3$ but it is not continuous at $x = 3$. (Other examples are possible.)

CHAPTER TWO

Solutions for Section 2.1

Exercises

1. For t between 2 and 5, we have

$$\text{Average velocity} = \frac{\Delta s}{\Delta t} = \frac{400 - 135}{5 - 2} = \frac{265}{3} \text{ km/hr.}$$

The average velocity on this part of the trip was $265/3$ km/hr.

5. Using $h = 0.1, 0.01, 0.001$, we see

$$\frac{(3 + 0.1)^3 - 27}{0.1} = 27.91$$
$$\frac{(3 + 0.01)^3 - 27}{0.01} = 27.09$$
$$\frac{(3 + 0.001)^3 - 27}{0.001} = 27.009.$$

These calculations suggest that $\lim\limits_{h \to 0} \dfrac{(3 + h)^3 - 27}{h} = 27$.

9. For $-0.5 \le \theta \le 0.5$, $0 \le y \le 3$, the graph of $y = \dfrac{\sin(2\,\theta)}{\theta}$ is shown in Figure 2.1. Therefore, $\lim\limits_{\theta \to 0} \dfrac{\sin(2\,\theta)}{\theta} = 2$.

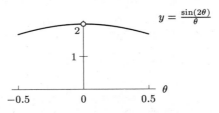

Figure 2.1

Problems

13.

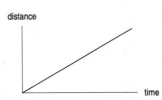

17. Between 1804 and 1927, the world's population increased 1 billion people in 123 years, for an average rate of change of $1/123$ billion people per year. We convert this to people per minute:

$$\frac{1,000,000,000}{123} \text{ people/year} \cdot \frac{1}{60 \cdot 24 \cdot 365} \text{ years/minute} = 15.47 \text{ people/minute.}$$

Between 1804 and 1927, the population of the world increased at an average rate of 15.47 people per minute. Similarly, we find the following:
Between 1927 and 1960, the increase was 57.65 people per minute.
Between 1960 and 1974, the increase was 135.90 people per minute.
Between 1974 and 1987, the increase was 146.35 people per minute.
Between 1987 and 1999, the increase was 158.55 people per minute.

Solutions for Section 2.2

Exercises

1. (a) As x approaches -2 from either side, the values of $f(x)$ get closer and closer to 3, so the limit appears to be about 3.

(b) As x approaches 0 from either side, the values of $f(x)$ get closer and closer to 7. (Recall that to find a limit, we are interested in what happens to the function near x but not at x.) The limit appears to be about 7.

(c) As x approaches 2 from either side, the values of $f(x)$ get closer and closer to 3 on one side of $x = 2$ and get closer and closer to 2 on the other side of $x = 2$. Thus the limit does not exist.

(d) As x approaches 4 from either side, the values of $f(x)$ get closer and closer to 8. (Again, recall that we don't care what happens right at $x = 4$.) The limit appears to be about 8.

5. From Table 2.1, it appears the limit is 0. This is confirmed by Figure 2.2. An appropriate window is $-0.005 < x < 0.005$, $-0.01 < y < 0.01$.

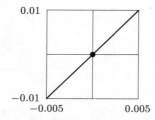

Figure 2.2

Table 2.1

x	$f(x)$
0.1	0.1987
0.01	0.0200
0.001	0.0020
0.0001	0.0002

x	$f(x)$
-0.0001	-0.0002
-0.001	-0.0020
-0.01	-0.0200
-0.1	-0.1987

9. From Table 2.2, it appears the limit is 1. This is confirmed by Figure 2.3. An appropriate window is $-0.0198 < x < 0.0198$, $0.99 < y < 1.01$.

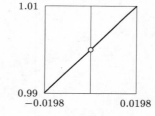

Figure 2.3

Table 2.2

x	$f(x)$
0.1	1.0517
0.01	1.0050
0.001	1.0005
0.0001	1.0001

x	$f(x)$
-0.0001	1.0000
-0.001	0.9995
-0.01	0.9950
-0.1	0.9516

13. From Table 2.3, it appears the limit is 0. Figure 2.4 confirms this. An appropriate window is $1.55 < x < 1.59$, $-0.01 < y < 0.01$.

Table 2.3

x	$f(x)$
1.6708	-0.0500
1.5808	-0.0050
1.5718	-0.0005
1.5709	-0.0001
1.5707	0.0001
1.5698	0.0005
1.5608	0.0050
1.4708	0.0500

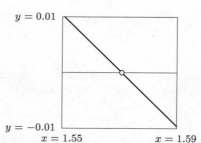

Figure 2.4

17. $\lim_{h \to 0} \dfrac{1}{h} \left(\dfrac{1}{(1+h)^2} - 1 \right) = \lim_{h \to 0} \dfrac{1 - (1 + 2h + h^2)}{h(1+h)^2} = \lim_{h \to 0} \dfrac{-2 - h}{(1+h)^2} = -2$

21. $f(x) = \dfrac{|x-2|}{x} = \begin{cases} \dfrac{x-2}{x}, & x > 2 \\[2mm] -\dfrac{x-2}{x}, & x < 2 \end{cases}$

Figure 2.5 confirms that $\lim\limits_{x \to 2^+} f(x) = \lim\limits_{x \to 2^-} f(x) = \lim\limits_{x \to 2} f(x) = 0.$

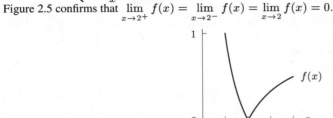

Figure 2.5

Problems

25. The only change is that, instead of considering all x near c, we only consider x near to and greater than c. Thus the phrase "$|x - c| < \delta$" must be replaced by "$c < x < c + \delta$." Thus, we define

$$\lim_{x \to c^+} f(x) = L$$

to mean that for any $\epsilon > 0$ (as small as we want), there is a $\delta > 0$ (sufficiently small) such that if $c < x < c + \delta$, then $|f(x) - L| < \epsilon$.

29. We use values of h approaching, but not equal to, zero. If we let $h = 0.01, 0.001, 0.0001, 0.00001$, we calculate the values $2.7048, 2.7169, 2.7181$, and 2.7183. If we let $h = -0.01\ -0.001, -0.0001, -0.00001$, we get values $2.7320, 2.7196, 2.7184$, and 2.7183. These numbers suggest that the limit is the number $e = 2.71828\ldots$. However, these calculations cannot tell us that the limit is exactly e; for that a proof is needed.

33. Divide numerator and denominator by x^3, giving

$$f(x) = \frac{2x^3 - 16x^2}{4x^2 + 3x^3} = \frac{2 - 16/x}{4/x + 3},$$

so

$$\lim_{x \to \infty} f(x) = \lim_{x \to \infty} \frac{2 - 16/x}{4/x + 3} = \frac{\lim_{x \to \infty}(2 - 16/x)}{\lim_{x \to \infty}(4/x + 3)} = \frac{2}{3}.$$

37. Because the denominator equals 0 when $x = 4$, so must the numerator. This means $k^2 = 16$ and the choices for k are 4 or -4.

41. For the numerator, $\lim\limits_{x \to -\infty} \left(e^{2x} - 5\right) = -5$. If $k > 0$, $\lim\limits_{x \to -\infty} \left(e^{kx} + 3\right) = 3$, so the quotient has a limit of $-5/3$. If $k = 0$, $\lim\limits_{x \to -\infty} \left(e^{kx} + 3\right) = 4$, so the quotient has limit of $-5/4$. If $k < 0$, the limit of the quotient is given by $\lim\limits_{x \to -\infty} \left(e^{2x} - 5\right)/\left(e^{kx} + 3\right) = 0$.

45. **(a)** Since $\sin(n\pi) = 0$ for $n = 1, 2, 3, \ldots$ the sequence of x-values

$$\frac{1}{\pi}, \frac{1}{2\pi}, \frac{1}{3\pi}, \ldots$$

works. These x-values $\to 0$ and are zeroes of $f(x)$.

(b) Since $\sin(n\pi/2) = 1$ for $n = 1, 5, 9 \ldots$ the sequence of x-values

$$\frac{2}{\pi}, \frac{2}{5\pi}, \frac{2}{9\pi}, \ldots$$

works.

(c) Since $\sin(n\pi)/2 = -1$ for $n = 3, 7, 11, \ldots$ the sequence of x-values

$$\frac{2}{3\pi}, \frac{2}{7\pi}, \frac{2}{11\pi} \ldots$$

works.

(d) Any two of these sequences of x-values show that if the limit were to exist, then it would have to have two (different) values: 0 and 1, or 0 and -1, or 1 and -1. Hence, the limit can not exist.

Solutions for Section 2.3

Exercises

1. The derivative, $f'(2)$, is the rate of change of x^3 at $x = 2$. Notice that each time x changes by 0.001 in the table, the value of x^3 changes by 0.012. Therefore, we estimate

$$f'(2) = \frac{\text{Rate of change}}{\text{of } f \text{ at } x = 2} \approx \frac{0.012}{0.001} = 12.$$

The function values in the table look exactly linear because they have been rounded. For example, the exact value of x^3 when $x = 2.001$ is 8.012006001, not 8.012. Thus, the table can tell us only that the derivative is approximately 12. Example 5 on page 82 shows how to compute the derivative of $f(x)$ exactly.

5. $f'(1) = \lim\limits_{h \to 0} \dfrac{\log(1 + h) - \log 1}{h} = \lim\limits_{h \to 0} \dfrac{\log(1 + h)}{h}$

Evaluating $\frac{\log(1+h)}{h}$ for $h = 0.01, 0.001$, and 0.0001, we get $0.43214, 0.43408, 0.43427$, so $f'(1) \approx 0.43427$. The corresponding secant lines are getting steeper, because the graph of $\log x$ is concave down. We thus expect the limit to be more than 0.43427. If we consider negative values of h, the estimates are too large. We can also see this from the graph below:

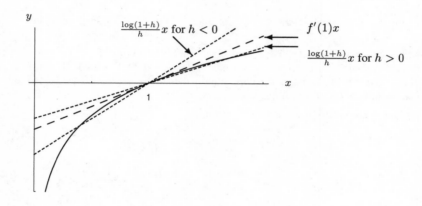

9. (a) The average rate of change from $x = a$ to $x = b$ is the slope of the line between the points on the curve with $x = a$ and $x = b$. Since the curve is concave down, the line from $x = 1$ to $x = 3$ has a greater slope than the line from $x = 3$ to $x = 5$, and so the average rate of change between $x = 1$ and $x = 3$ is greater than that between $x = 3$ and $x = 5$.
   (b) Since f is increasing, $f(5)$ is the greater.
   (c) As in part (a), f is concave down and f' is decreasing throughout so $f'(1)$ is the greater.

13.

$$f'(1) = \lim_{h \to 0} \frac{f(1 + h) - f(1)}{h} = \lim_{h \to 0} \frac{((1 + h)^3 + 5) - (1^3 + 5)}{h}$$

$$= \lim_{h \to 0} \frac{1 + 3h + 3h^2 + h^3 + 5 - 1 - 5}{h} = \lim_{h \to 0} \frac{3h + 3h^2 + h^3}{h}$$

$$= \lim_{h \to 0} (3 + 3h + h^2) = 3.$$

17. As we saw in the answer to Problem 11, the slope of the tangent line to $f(x) = x^3$ at $x = -2$ is 12. When $x = -2$, $f(x) = -8$ so we know the point $(-2, -8)$ is on the tangent line. Thus the equation of the tangent line is $y = 12(x + 2) - 8 = 12x + 16$.

Problems

21. The coordinates of A are $(4, 25)$. See Figure 2.6. The coordinates of B and C are obtained using the slope of the tangent line. Since $f'(4) = 1.5$, the slope is 1.5

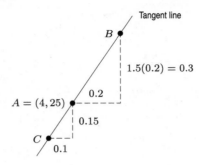

Figure 2.6

From A to B, $\Delta x = 0.2$, so $\Delta y = 1.5(0.2) = 0.3$. Thus, at C we have $y = 25 + 0.3 = 25.3$. The coordinates of B are $(4.2, 25.3)$.

From A to C, $\Delta x = -0.1$, so $\Delta y = 1.5(-0.1) = -0.15$. Thus, at C we have $y = 25 - 0.15 = 24.85$. The coordinates of C are $(3.9, 24.85)$.

25. Figure 2.7 shows the quantities in which we are interested.

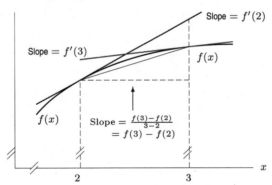

Figure 2.7

The quantities $f'(2)$, $f'(3)$ and $f(3) - f(2)$ have the following interpretations:

- $f'(2) = $ slope of the tangent line at $x = 2$
- $f'(3) = $ slope of the tangent line at $x = 3$
- $f(3) - f(2) = \frac{f(3)-f(2)}{3-2} = $ slope of the secant line from $f(2)$ to $f(3)$.

From Figure 2.7, it is clear that $0 < f(3) - f(2) < f'(2)$. By extending the secant line past the point $(3, f(3))$, we can see that it lies above the tangent line at $x = 3$.

Thus

$$0 < f'(3) < f(3) - f(2) < f'(2).$$

29. (a)

$$f'(0) = \lim_{h \to 0} \frac{\overbrace{\sin h}^{h \text{ in degrees}} - \overbrace{\sin 0}^{0}}{h} = \frac{\sin h}{h}.$$

To four decimal places,

$$\frac{\sin 0.2}{0.2} \approx \frac{\sin 0.1}{0.1} \approx \frac{\sin 0.01}{0.01} \approx \frac{\sin 0.001}{0.001} \approx 0.01745$$

so $f'(0) \approx 0.01745$.

(b) Consider the ratio $\frac{\sin h}{h}$. As we approach 0, the numerator, $\sin h$, will be much smaller in magnitude if h is in degrees than it would be if h were in radians. For example, if $h = 1°$ radian, $\sin h = 0.8415$, but if $h = 1$ degree, $\sin h = 0.01745$. Thus, since the numerator is smaller for h measured in degrees while the denominator is the same, we expect the ratio $\frac{\sin h}{h}$ to be smaller.

33. We want to approximate $P'(0)$ and $P'(2)$. Since for small h

$$P'(0) \approx \frac{P(h) - P(0)}{h},$$

if we take $h = 0.01$, we get

$$P'(0) \approx \frac{1.15(1.014)^{0.01} - 1.15}{0.01} = 0.01599 \text{ billion/year}$$

$$= 16.0 \text{ million people/year}$$

$$P'(2) \approx \frac{1.15(1.014)^{2.01} - 1.15(1.014)^2}{0.01} = 0.0164 \text{ billion/year}$$

$$= 16.4 \text{ million people/year}$$

Solutions for Section 2.4

Exercises

1. The graph is that of the line $y = -2x + 2$. The slope, and hence the derivative, is -2.

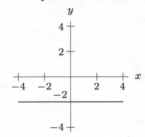

5.

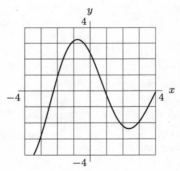

9.

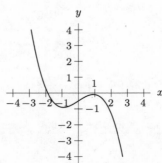

13. Since $1/x = x^{-1}$, using the power rule gives

$$\frac{d}{dx}(x^{-1}) = (-1)x^{-2} = -\frac{1}{x^2}.$$

Using the definition of the derivative, we have

$$k'(x) = \lim_{h \to 0} \frac{k(x+h) - k(x)}{h} = \lim_{h \to 0} \frac{\frac{1}{x+h} - \frac{1}{x}}{h} = \lim_{h \to 0} \frac{x - (x+h)}{h(x+h)x}$$

$$= \lim_{h \to 0} \frac{-h}{h(x+h)x} = \lim_{h \to 0} \frac{-1}{(x+h)x} = -\frac{1}{x^2}.$$

17.

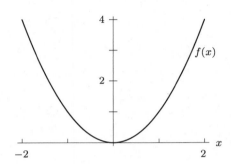
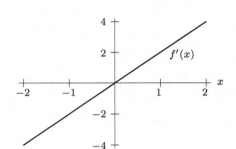

Problems

21. We know that $f'(x) \approx \dfrac{f(x+h) - f(x)}{h}$. For this problem, we'll take the average of the values obtained for $h = 1$ and $h = -1$; that's the average of $f(x+1) - f(x)$ and $f(x) - f(x-1)$ which equals $\dfrac{f(x+1) - f(x-1)}{2}$. Thus,

$f'(0) \approx f(1) - f(0) = 13 - 18 = -5.$
$f'(1) \approx [f(2) - f(0)]/2 = [10 - 18]/2 = -4.$
$f'(2) \approx [f(3) - f(1)]/2 = [9 - 13]/2 = -2.$
$f'(3) \approx [f(4) - f(2)]/2 = [9 - 10]/2 = -0.5.$
$f'(4) \approx [f(5) - f(3)]/2 = [11 - 9]/2 = 1.$
$f'(5) \approx [f(6) - f(4)]/2 = [15 - 9]/2 = 3.$
$f'(6) \approx [f(7) - f(5)]/2 = [21 - 11]/2 = 5.$
$f'(7) \approx [f(8) - f(6)]/2 = [30 - 15]/2 = 7.5.$
$f'(8) \approx f(8) - f(7) = 30 - 21 = 9.$
The rate of change of $f(x)$ is positive for $4 \le x \le 8$, negative for $0 \le x \le 3$. The rate of change is greatest at about $x = 8$.

25. This function is decreasing for $x < 2$ and increasing for $x > 2$ and so the derivative is negative for $x < 2$ and positive for $x > 2$. One possible graph is shown in Figure 2.8.

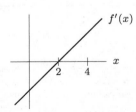

Figure 2.8

29.

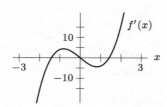

33. From the given information we know that f is increasing for values of x less than -2, is decreasing between $x = -2$ and $x = 2$, and is constant for $x > 2$. Figure 2.9 shows a possible graph—yours may be different.

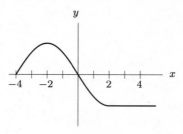

Figure 2.9

37. (a) Graph II
   (b) Graph I
   (c) Graph III

41. If $f(x)$ is even, its graph is symmetric about the y-axis. So the tangent line to f at $x = x_0$ is the same as that at $x = -x_0$ reflected about the y-axis.

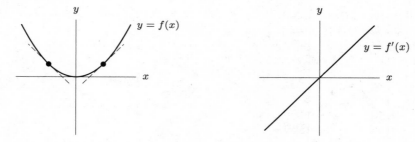

So the slopes of these two tangent lines are opposite in sign, so $f'(x_0) = -f'(-x_0)$, and f' is odd.

Solutions for Section 2.5

Exercises

1. (a) As the cup of coffee cools, the temperature decreases, so $f'(t)$ is negative.
   (b) Since $f'(t) = dH/dt$, the units are degrees Celsius per minute. The quantity $f'(20)$ represents the rate at which the coffee is cooling, in degrees per minute, 20 minutes after the cup is put on the counter.

5. Since B is measured in dollars and t is measured in years, dB/dt is measured in dollars per year. We can interpret dB as the extra money added to your balance in dt years. Therefore dB/dt represents how fast your balance is growing, in units of dollars/year.

9. The units of $f'(x)$ are feet/mile. The derivative, $f'(x)$, represents the rate of change of elevation with distance from the source, so if the river is flowing downhill everywhere, the elevation is always decreasing and $f'(x)$ is always negative. (In fact, there may be some stretches where the elevation is more or less constant, so $f'(x) = 0$.)

Problems

13. Since $f(t) = 1.15(1.014)^t$, we have
$$f(6) = 1.15(1.014)^6 = 1.25.$$
To estimate $f'(6)$, we use a small interval around 6:
$$f'(6) \approx \frac{f(6.001) - f(6)}{6.001 - 6} = \frac{1.15(1.014)^{6.001} - 1.15(1.014)^6}{0.001} = 0.0174.$$
We see that $f(6) = 1.25$ billion people and $f'(6) = 0.0174$ billion people per year. This model tells us that the population of China was about 1,250,000,000 people in 1999 and was growing at a rate of about 17,400,000 people per year at that time.

17. Units of $g'(55)$ are mpg/mph. The statement $g'(55) = -0.54$ means that at 55 miles per hour the fuel efficiency (in miles per gallon, or mpg) of the car decreases at a rate of approximately one half mpg as the velocity increases by one mph.

21. (a) The units of compliance are units of volume per units of pressure, or liters per centimeter of water.
   (b) The increase in volume for a 5 cm reduction in pressure is largest between 10 and 15 cm. Thus, the compliance appears maximum between 10 and 15 cm of pressure reduction. The derivative is given by the slope, so
$$\text{Compliance} \approx \frac{0.70 - 0.49}{15 - 10} = 0.042 \text{ liters per centimeter.}$$
   (c) When the lung is nearly full, it cannot expand much more to accommodate more air.

Solutions for Section 2.6

Exercises

1. (a) Since the graph is below the x-axis at $x = 2$, $f(2)$ is negative.
   (b) Since $f(x)$ is decreasing at $x = 2$, $f'(2)$ is negative.
   (c) Since $f(x)$ is concave up at $x = 2$, $f''(2)$ is positive.

5. The function is everywhere increasing and concave up. One possible graph is shown in Figure 2.10.

Figure 2.10

9. $f'(x) = 0$
   $f''(x) = 0$

13. $f'(x) < 0$
   $f''(x) < 0$

Problems

17. (a) $dP/dt > 0$ and $d^2P/dt^2 > 0$.

   (b) $dP/dt < 0$ and $d^2P/dt^2 > 0$ (but dP/dt is close to zero).

21. (a) The EPA will say that the rate of discharge is still rising. The industry will say that the rate of discharge is increasing less quickly, and may soon level off or even start to fall.
   (b) The EPA will say that the rate at which pollutants are being discharged is levelling off, but not to zero — so pollutants will continue to be dumped in the lake. The industry will say that the rate of discharge has decreased significantly.

Solutions for Section 2.7

Exercises

1. **(a)** Function f is not continuous at $x = 1$.
  (b) Function f appears not differentiable at $x = 1, 2, 3$.

5. Yes.

Problems

9. We can see from Figure 2.11 that the graph of f oscillates infinitely often between the curves $y = x^2$ and $y = -x^2$ near the origin. Thus the slope of the line from $(0, 0)$ to $(h, f(h))$ oscillates between h (when $f(h) = h^2$ and $\frac{f(h)-0}{h-0} = h$) and $-h$ (when $f(h) = -h^2$ and $\frac{f(h)-0}{h-0} = -h$) as h tends to zero. So, the limit of the slope as h tends to zero is 0, which is the derivative of f at the origin. Another way to see this is to observe that

$$\lim_{h \to 0} \frac{f(h) - f(0)}{h} = \lim_{h \to 0} \left(\frac{h^2 \sin(\frac{1}{h})}{h} \right)$$
$$= \lim_{h \to 0} h \sin(\frac{1}{h})$$
$$= 0,$$

since $\lim_{h \to 0} h = 0$ and $-1 \leq \sin(\frac{1}{h}) \leq 1$ for any h. Thus f is differentiable at $x = 0$, and $f'(0) = 0$.

Figure 2.11

13. **(a)** Since

$$\lim_{r \to r_0^-} E = kr_0$$

and

$$\lim_{r \to r_0^+} E = \frac{kr_0^2}{r_0} = kr_0$$

and

$$E(r_0) = kr_0,$$

we see that E is continuous at r_0.
  (b) The function E is not differentiable at $r = r_0$ because the graph has a corner there. The slope is positive for $r < r_0$ and the slope is negative for $r > r_0$.
  (c)

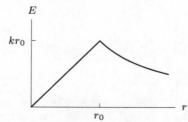

17. Since $f(x) = x$ is continuous, Theorem 2.2 on page 95 shows that products of the form $f(x) \cdot f(x) = x^2$ and $f(x) \cdot x^2 = x^3$, etc., are continuous. By a similar argument, x^n is continuous for any $n > 0$.

Solutions for Chapter 2 Review

Exercises

1.

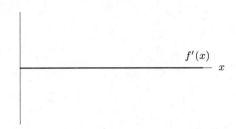

5.

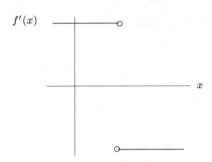

9. From Table 2.4, it appears the limit is 0. This is confirmed by Figure 2.12. An appropriate window is $-0.015 < x < 0.015$, $-0.01 < y < 0.01$.

Table 2.4

x	$f(x)$		x	$f(x)$
0.1	0.0666		-0.0001	-0.0001
0.01	0.0067		-0.001	-0.0007
0.001	0.0007		-0.01	-0.0067
0.0001	0		-0.1	-0.0666

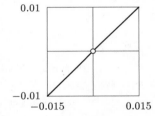

Figure 2.12

13. $\displaystyle \lim_{h \to 0} \frac{(a+h)^2 - a^2}{h} = \lim_{h \to 0} \frac{a^2 + 2ah + h^2 - a^2}{h} = \lim_{h \to 0} (2a + h) = 2a$

17. We combine terms in the numerator and multiply top and bottom by $\sqrt{a} + \sqrt{a+h}$.

$$\frac{1}{\sqrt{a+h}} - \frac{1}{\sqrt{a}} = \frac{\sqrt{a} - \sqrt{a+h}}{\sqrt{a+h}\sqrt{a}} = \frac{(\sqrt{a} - \sqrt{a+h})(\sqrt{a} + \sqrt{a+h})}{\sqrt{a+h}\sqrt{a}(\sqrt{a} + \sqrt{a+h})}$$

$$= \frac{a - (a+h)}{\sqrt{a+h}\sqrt{a}(\sqrt{a} + \sqrt{a+h})}$$

Therefore $\displaystyle \lim_{h \to 0} \frac{1}{h}\left(\frac{1}{\sqrt{a+h}} - \frac{1}{\sqrt{a}} \right) = \lim_{h \to 0} \frac{-1}{\sqrt{a+h}\sqrt{a}(\sqrt{a} + \sqrt{a+h})} = \frac{-1}{2(\sqrt{a})^3}$

Problems

21. Since $f(2) = 3$ and $f'(2) = 1$, near $x = 2$ the graph looks like the segment shown in Figure 2.13.

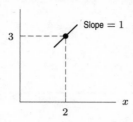

Figure 2.13

(a) If $f(x)$ is even, then the graph of $f(x)$ near $x = 2$ and $x = -2$ looks like Figure 2.14. Thus $f(-2) = 3$ and $f'(-2) = -1$.

(b) If $f(x)$ is odd, then the graph of $f(x)$ near $x = 2$ and $x = -2$ looks like Figure 2.15. Thus $f(-2) = -3$ and $f'(-2) = 1$.

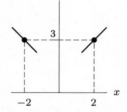

Figure 2.14: For f even

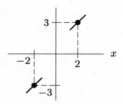

Figure 2.15: For f odd

25. (a) Since the point $A = (7, 3)$ is on the graph of f, we have $f(7) = 3$.

(b) The slope of the tangent line touching the curve at $x = 7$ is given by

$$\text{Slope} = \frac{\text{Rise}}{\text{Run}} = \frac{3.8 - 3}{7.2 - 7} = \frac{0.8}{0.2} = 4.$$

Thus, $f'(7) = 4$.

29. $f(10) = 240,000$ means that if the commodity costs \$10, then 240,000 units of it will be sold. $f'(10) = -29,000$ means that if the commodity costs \$10 now, each \$1 increase in price will cause a decline in sales of 29,000 units.

33. By tracing on a calculator or solving equations, we find the following values of δ:
For $\epsilon = 0.1$, $\delta \le 0.45$.
For $\epsilon = 0.001$, $\delta \le 0.0447$.
For $\epsilon = 0.00001$, $\delta \le 0.00447$.

37. (a) The population varies periodically with a period of 12 months (i.e. one year).

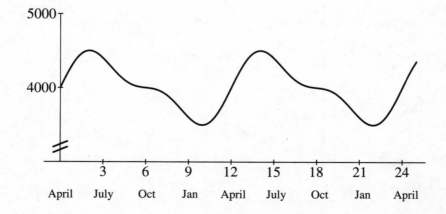

(b) The herd is largest about June 1st when there are about 4500 deer.

(c) The herd is smallest about February 1st when there are about 3500 deer.

(d) The herd grows the fastest about April 1st. The herd shrinks the fastest about July 15 and again about December 15.

(e) It grows the fastest about April 1st when the rate of growth is about 400 deer/month, i.e about 13 new fawns per day.

41. (a)

Table 2.5

x	$\frac{\sinh(x+0.001)-\sinh(x)}{0.001}$	$\frac{\sinh(x+0.0001)-\sinh(x)}{0.0001}$	so $f'(0) \approx$	$\cosh(x)$
0	1.00000	1.00000	1.00000	1.00000
0.3	1.04549	1.04535	1.04535	1.04534
0.7	1.25555	1.25521	1.25521	1.25517
1	1.54367	1.54314	1.54314	1.54308

(b) It seems that they are approximately the same, i.e. the derivative of $\sinh(x) = \cosh(x)$ for $x = 0, 0.3, 0.7$, and 1.

CAS Challenge Problems

45. (a) The CAS gives the same derivative, $1/x$, in all three cases.

(b) From the properties of logarithms, $g(x) = \ln(2x) = \ln 2 + \ln x = f(x) + \ln 2$. So the graph of g is the same shape as the graph of f, only shifted up by $\ln 2$. So the graphs have the same slope everywhere, and therefore the two functions have the same derivative. By the same reasoning, $h(x) = f(x) + \ln 3$, so h and f have the same derivative as well.

CHECK YOUR UNDERSTANDING

1. False. For example, the car could slow down or even stop at one minute after 2 pm, and then speed back up to 60 mph at one minute before 3 pm. In this case the car would travel only a few miles during the hour, much less than 50 miles.

5. True. By definition, Average velocity = Distance traveled/Time.

9. True. The derivative of a function is the limit of difference quotients. A few difference quotients can be computed from the table, but the limit can not be computed from the table.

13. True. Shifting a graph vertically does not change the shape of the graph and so it does not change the slopes of the tangent lines to the graph.

17. True. The second derivative $f''(x)$ is the derivative of $f'(x)$. Thus the derivative of $f'(x)$ is positive, and so $f'(x)$ is increasing.

21. True. Let $f(x) = |x - 3|$. Then $f(x)$ is continuous for all x but not differentiable at $x = 3$ because its graph has a corner there. Other answers are possible.

25. False. For example, $f(x) = |x|$ is not differentiable at $x = 0$, but it is continuous at $x = 0$.

29. True, by Property 2 of limits in Theorem 2.1.

33. True. Suppose instead that $\lim_{x \to 3} g(x)$ does not exist but $\lim_{x \to 3}(f(x)g(x))$ did exist. Since $\lim_{x \to 3} f(x)$ exists and is not zero, then $\lim_{x \to 3}((f(x)g(x))/f(x))$ exists, by Property 4 of limits in Theorem 2.1. Furthermore, $f(x) \neq 0$ for all x in some interval about 3, so $(f(x)g(x))/f(x) = g(x)$ for all x in that interval. Thus $\lim_{x \to 3} g(x)$ exists. This contradicts our assumption that $\lim_{x \to 3} g(x)$ does not exist.

37. False. Although x may be far from c, the value of $f(x)$ could be close to L. For example, suppose $f(x) = L$, the constant function.

CHAPTER THREE

Solutions for Section 3.1

Exercises

1. The derivative, $f'(x)$, is defined as

$$f'(x) = \lim_{h \to 0} \frac{f(x+h) - f(x)}{h}.$$

If $f(x) = 7$, then

$$f'(x) = \lim_{h \to 0} \frac{7 - 7}{h} = \lim_{h \to 0} \frac{0}{h} = 0.$$

5. $y' = 11x^{-12}$.

9. $y' = \frac{3}{4}x^{-1/4}$.

13. $f'(x) = ex^{e-1}$.

17. Dividing gives $g(t) = t^2 + k/t$ so $g'(t) = 2t - \dfrac{k}{t^2}$.

21. $y' = 15t^4 - \frac{5}{2}t^{-1/2} - \frac{7}{t^2}$.

25. $f(z) = \dfrac{z}{3} + \dfrac{1}{3}z^{-1} = \dfrac{1}{3}\left(z + z^{-1}\right)$, so $f'(z) = \dfrac{1}{3}\left(1 - z^{-2}\right) = \dfrac{1}{3}\left(\dfrac{z^2 - 1}{z^2}\right)$.

29. Since $4/3$, π, and b are all constants, we have

$$\frac{dV}{dr} = \frac{4}{3}\pi(2r)b = \frac{8}{3}\pi rb.$$

33. $g'(x) = -\dfrac{1}{2}(5x^4 + 2)$.

Problems

37. The x is in the exponent and we haven't learned how to handle that yet.

41. $y' = -2/3z^3$.   (power rule and sum rule)

45. The graph increases when $dy/dx > 0$:

$$\frac{dy}{dx} = 5x^4 - 5 > 0$$
$$5(x^4 - 1) > 0 \quad \text{so} \quad x^4 > 1 \quad \text{so} \quad x > 1 \text{ or } x < -1.$$

The graph is concave up when $d^2y/dx^2 > 0$:

$$\frac{d^2y}{dx^2} = 20x^3 > 0 \quad \text{so} \quad x > 0.$$

We need values of x where $\{x > 1 \text{ or } x < -1\}$ AND $\{x > 0\}$, which implies $x > 1$. Thus, both conditions hold for all values of x larger than 1.

49. Differentiating gives

$$f'(x) = 6x^2 - 4x \quad \text{so} \quad f'(1) = 6 - 4 = 2.$$

Thus the equation of the tangent line is $(y - 1) = 2(x - 1)$ or $y = 2x - 1$.

53. Since $f(x) = ax^n$, $f'(x) = anx^{n-1}$. We know that $f'(2) = (an)2^{n-1} = 3$, and $f'(4) = (an)4^{n-1} = 24$. Therefore,

$$\frac{f'(4)}{f'(2)} = \frac{24}{3}$$
$$\frac{(an)4^{n-1}}{(an)2^{n-1}} = \left(\frac{4}{2}\right)^{n-1} = 8$$
$$2^{n-1} = 8, \text{ and thus } n = 4.$$

Substituting $n = 4$ into the expression for $f'(2)$, we get $3 = a(4)(8)$, or $a = 3/32$.

57. (a) The average velocity between $t = 0$ and $t = 2$ is given by

$$\text{Average velocity} = \frac{f(2) - f(0)}{2 - 0} = \frac{-4.9(2^2) + 25(2) + 3 - 3}{2 - 0} = \frac{33.4 - 3}{2} = 15.2 \text{ m/sec.}$$

(b) Since $f'(t) = -9.8t + 25$, we have

$$\text{Instantaneous velocity} = f'(2) = -9.8(2) + 25 = 5.4 \text{ m/sec.}$$

(c) Acceleration is given $f''(t) = -9.8$. The acceleration at $t = 2$ (and all other times) is the acceleration due to gravity, which is -9.8 m/sec^2.

(d) We can use a graph of height against time to estimate the maximum height of the tomato. See Figure 3.1. Alternately, we can find the answer analytically. The maximum height occurs when the velocity is zero and $v(t) = -9.8t + 25 = 0$ when $t = 2.6$ sec. At this time the tomato is at a height of $f(2.6) = 34.9$. The maximum height is 34.9 meters.

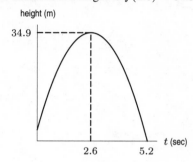

Figure 3.1

(e) We see in Figure 3.1 that the tomato hits ground at about $t = 5.2$ seconds. Alternately, we can find the answer analytically. The tomato hits the ground when

$$f(t) = -4.9t^2 + 25t + 3 = 0.$$

We solve for t using the quadratic formula:

$$t = \frac{-25 \pm \sqrt{(25)^2 - 4(-4.9)(3)}}{2(-4.9)}$$

$$t = \frac{-25 \pm \sqrt{683.8}}{-9.8}$$

$$t = -0.12 \quad \text{and} \quad t = 5.2.$$

We use the positive values, so the tomato hits the ground at $t = 5.2$ seconds.

61. $V = \frac{4}{3}\pi r^3$. Differentiating gives $\frac{dV}{dr} = 4\pi r^2 = $ surface area of a sphere.

The difference quotient $\frac{V(r+h) - V(r)}{h}$ is the volume between two spheres divided by the change in radius. Furthermore, when h is very small, the difference between volumes, $V(r + h) - V(r)$, is like a coating of paint of depth h applied to the surface of the sphere. The volume of the paint is about $h \cdot$ (Surface Area) for small h: dividing by h gives back the surface area.

Thinking about the derivative as the rate of change of the function for a small change in the variable gives another way of seeing the result. If you increase the radius of a sphere a small amount, the volume increases by a very thin layer whose volume is the surface area at that radius multiplied by that small amount.

Solutions for Section 3.2

Exercises

1. $f'(x) = 2e^x + 2x$.

5. $y' = 10x + (\ln 2)2^x$.

9. $\frac{dy}{dx} = \frac{1}{3}(\ln 3)3^x - \frac{33}{2}(x^{-\frac{3}{2}})$.

13. $y = e^\theta e^{-1} \quad y' = \frac{d}{d\theta}(e^\theta e^{-1}) = e^{-1}\frac{d}{d\theta}e^\theta = e^\theta e^{-1} = e^{\theta - 1}$.

17. $f'(x) = 3x^2 + 3^x \ln 3$

21. $f'(x) = (\ln \pi)\pi^x$.

25. $f'(z) = (2 \ln 3)z + (\ln 4)e^z$.

29. We can take the derivative of the sum $x^2 + 2^x$, but not the product.

33. The exponent is x^2, and we haven't learned what to do about that yet.

Problems

37. Since $P = 1 \cdot (1.05)^t$, $\frac{dP}{dt} = \ln(1.05)1.05^t$. When $t = 10$,

$$\frac{dP}{dt} = (\ln 1.05)(1.05)^{10} \approx \$0.07947/\text{year} \approx 7.95\text{¢}/\text{year}.$$

41. (a) The rate of change of the population is $P'(t)$. If $P'(t)$ is proportional to $P(t)$, we have

$$P'(t) = kP(t).$$

(b) If $P(t) = Ae^{kt}$, then $P'(t) = kAe^{kt} = kP(t)$.

45. The derivative of e^x is $\frac{d}{dx}(e^x) = e^x$. Thus the tangent line at $x = 0$, has slope $e^0 = 1$, and the tangent line is $y = x + 1$. A function which is always concave up will always stay above any of its tangent lines. Thus $e^x \geq x + 1$ for all x, as shown in the figure below.

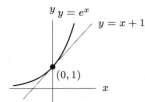

Solutions for Section 3.3

Exercises

1. By the product rule, $f'(x) = 2x(x^3 + 5) + x^2(3x^2) = 2x^4 + 3x^4 + 10x = 5x^4 + 10x$. Alternatively, $f'(x) = (x^5 + 5x^2)' = 5x^4 + 10x$. The two answers should, and do, match.

5. $y' = \frac{1}{2\sqrt{x}}2^x + \sqrt{x}(\ln 2)2^x$.

9. $y' = (3t^2 - 14t)e^t + (t^3 - 7t^2 + 1)e^t = (t^3 - 4t^2 - 14t + 1)e^t$.

13. $q'(r) = \frac{3(5r + 2) - 3r(5)}{(5r + 2)^2} = \frac{15r + 6 - 15r}{(5r + 2)^2} = \frac{6}{(5r + 2)^2}$

17. Using the quotient rule gives $\frac{dz}{dt} = \frac{(2t + 3)(t + 1) - (t^2 + 3t + 1)}{(t + 1)^2}$ or $\frac{dz}{dt} = \frac{t^2 + 2t + 2}{(t + 1)^2}$.

21. $\frac{d}{dz}\left(\frac{z^2 + 1}{\sqrt{z}}\right) = \frac{d}{dz}(z^{\frac{3}{2}} + z^{-\frac{1}{2}}) = \frac{3}{2}z^{\frac{1}{2}} - \frac{1}{2}z^{-\frac{3}{2}} = \frac{\sqrt{z}}{2}(3 - z^{-2})$.

25. $w'(x) = \frac{17e^x(2^x) - (\ln 2)(17e^x)2^x}{2^{2x}} = \frac{17e^x(2^x)(1 - \ln 2)}{2^{2x}} = \frac{17e^x(1 - \ln 2)}{2^x}$.

29. $w' = (3t^2 + 5)(t^2 - 7t + 2) + (t^3 + 5t)(2t - 7)$.

Problems

33. Since $f(0) = -5/1 = -5$, the tangent line passes through the point $(0, -5)$, so its vertical intercept is -5. To find the slope of the tangent line, we find the derivative of $f(x)$ using the quotient rule:

$$f'(x) = \frac{(x + 1) \cdot 2 - (2x - 5) \cdot 1}{(x + 1)^2} = \frac{7}{(x + 1)^2}.$$

At $x = 0$, the slope of the tangent line is $m = f'(0) = 7$. The equation of the tangent line is $y = 7x - 5$.

37. Since $\frac{d}{dx}e^{2x} = 2e^{2x}$ and $\frac{d}{dx}e^{3x} = 3e^{3x}$, we might guess that $\frac{d}{dx}e^{4x} = 4e^{4x}$.

41. (a) We have $h'(2) = f'(2) + g'(2) = 5 - 2 = 3$.
    (b) We have $h'(2) = f'(2)g(2) + f(2)g'(2) = 5(4) + 3(-2) = 14$.
    (c) We have $h'(2) = \dfrac{f'(2)g(2) - f(2)g'(2)}{(g(2))^2} = \dfrac{5(4) - 3(-2)}{4^2} = \dfrac{26}{16} = \dfrac{13}{8}$.

45. We want dR/dr_1. Solving for R:

$$\frac{1}{R} = \frac{1}{r_1} + \frac{1}{r_2} = \frac{r_2 + r_1}{r_1 r_2}, \quad \text{which gives } R = \frac{r_1 r_2}{r_2 + r_1}.$$

So, thinking of r_2 as a constant and using the quotient rule,

$$\frac{dR}{dr_1} = \frac{r_2(r_2 + r_1) - r_1 r_2(1)}{(r_2 + r_1)^2} = \frac{r_2^2}{(r_1 + r_2)^2}.$$

49. (a) $f'(x) = (x - 2) + (x - 1)$.
    (b) Think of f as the product of two factors, with the first as $(x - 1)(x - 2)$. (The reason for this is that we have already differentiated $(x - 1)(x - 2)$).

$$f(x) = [(x - 1)(x - 2)](x - 3).$$

Now $f'(x) = [(x - 1)(x - 2)]'(x - 3) + [(x - 1)(x - 2)](x - 3)'$
Using the result of a):

$$f'(x) = [(x - 2) + (x - 1)](x - 3) + [(x - 1)(x - 2)] \cdot 1$$
$$= (x - 2)(x - 3) + (x - 1)(x - 3) + (x - 1)(x - 2).$$

    (c) Because we have already differentiated $(x - 1)(x - 2)(x - 3)$, rewrite f as the product of two factors, the first being $(x - 1)(x - 2)(x - 3)$:

$$f(x) = [(x - 1)(x - 2)(x - 3)](x - 4)$$

Now $f'(x) = [(x - 1)(x - 2)(x - 3)]'(x - 4) + [(x - 1)(x - 2)(x - 3)](x - 4)'$.

$$f'(x) = [(x - 2)(x - 3) + (x - 1)(x - 3) + (x - 1)(x - 2)](x - 4)$$
$$+ [(x - 1)(x - 2)(x - 3)] \cdot 1$$
$$= (x - 2)(x - 3)(x - 4) + (x - 1)(x - 3)(x - 4)$$
$$+ (x - 1)(x - 2)(x - 4) + (x - 1)(x - 2)(x - 3).$$

From the solutions above, we can observe that when f is a product, its derivative is obtained by differentiating each factor in turn (leaving the other factors alone), and adding the results.

Solutions for Section 3.4

Exercises

1. $f'(x) = 99(x + 1)^{98} \cdot 1 = 99(x + 1)^{98}$.

5. $w' = 100(\sqrt{t} + 1)^{99} \left(\frac{1}{2\sqrt{t}} \right) = \frac{50}{\sqrt{t}}(\sqrt{t} + 1)^{99}$.

9. $g(x) = \pi e^{\pi x}$.

13. $k'(x) = 4(x^3 + e^x)^3 (3x^2 + e^x)$.

17. $\dfrac{d}{dt} e^{(1+3t)^2} = e^{(1+3t)^2} \dfrac{d}{dt}(1 + 3t)^2 = e^{(1+3t)^2} \cdot 2(1 + 3t) \cdot 3 = 6(1 + 3t)e^{(1+3t)^2}$.

21. $y' = \frac{3}{2} e^{\frac{3}{2}w}$.

25. $y' = 1 \cdot e^{-t^2} + te^{-t^2}(-2t)$

29. $f'(t) = 1 \cdot e^{5-2t} + te^{5-2t}(-2) = e^{5-2t}(1 - 2t)$.

33. $y' = \dfrac{-(3e^{3x} + 2x)}{(e^{3x} + x^2)^2}$.

37. $f'(\theta) = -1(1 + e^{-\theta})^{-2}(e^{-\theta})(-1) = \dfrac{e^{-\theta}}{(1 + e^{-\theta})^2}$.

41. $f(y) = \left[10^{(5-y)} \right]^{\frac{1}{2}} = 10^{\frac{5}{2} - \frac{1}{2}y}$
    $f'(y) = (\ln 10) \left(10^{\frac{5}{2} - \frac{1}{2}y} \right) \left(-\frac{1}{2} \right) = -\frac{1}{2}(\ln 10)(10^{\frac{5}{2} - \frac{1}{2}y})$.

45. Since a and b are constants, we have $f'(t) = ae^{bt}(b) = abe^{bt}$.

Problems

49. We have $f(2) = (2-1)^3 = 1$, so $(2, 1)$ is a point on the tangent line. Since $f'(x) = 3(x-1)^2$, the slope of the tangent line is

$$m = f'(2) = 3(2-1)^2 = 3.$$

The equation of the line is

$$y - 1 = 3(x - 2) \quad \text{or} \quad y = 3x - 5.$$

53. **(a)** $H(x) = F(G(x))$
$H(4) = F(G(4)) = F(2) = 1$
(b) $H(x) = F(G(x))$
$H'(x) = F'(G(x)) \cdot G'(x)$
$H'(4) = F'(G(4)) \cdot G'(4) = F'(2) \cdot 6 = 5 \cdot 6 = 30$
(c) $H(x) = G(F(x))$
$H(4) = G(F(4)) = G(3) = 4$
(d) $H(x) = G(F(x))$
$H'(x) = G'(F(x)) \cdot F'(x)$
$H'(4) = G'(F(4)) \cdot F'(4) = G'(3) \cdot 7 = 8 \cdot 7 = 56$
(e) $H(x) = \frac{F(x)}{G(x)}$
$H'(x) = \frac{G(x) \cdot F'(x) - F(x) \cdot G'(x)}{[G(x)]^2}$
$H'(4) = \frac{G(4) \cdot F'(4) - F(4) \cdot G'(4)}{[G(4)]^2} = \frac{2 \cdot 7 - 3 \cdot 6}{2^2} = \frac{14 - 18}{4} = \frac{-4}{4} = -1$

57. Yes. To see why, simply plug $x = \sqrt[3]{2t + 5}$ into the expression $3x^2 \dfrac{dx}{dt}$ and evaluate it. To do this, first we calculate $\dfrac{dx}{dt}$. By the chain rule,

$$\frac{dx}{dt} = \frac{d}{dt}(2t + 5)^{\frac{1}{3}} = \frac{2}{3}(2t + 5)^{-\frac{2}{3}} = \frac{2}{3}[(2t + 5)^{\frac{1}{3}}]^{-2}.$$

But since $x = (2t + 5)^{\frac{1}{3}}$, we have (by substitution)

$$\frac{dx}{dt} = \frac{2}{3}x^{-2}.$$

It follows that $3x^2 \dfrac{dx}{dt} = 3x^2 \left(\dfrac{2}{3}x^{-2}\right) = 2$.

61. We have $f(0) = 6$ and $f(10) = 6e^{0.013(10)} = 6.833$. The derivative of $f(t)$ is

$$f'(t) = 6e^{0.013t} \cdot 0.013 = 0.078e^{0.013t},$$

and so $f'(0) = 0.078$ and $f'(10) = 0.089$.

These values tell us that in 1999 (at $t = 0$), the population of the world was 6 billion people and the population was growing at a rate of 0.078 billion people per year. In the year 2009 (at $t = 10$), this model predicts that the population of the world will be 6.833 billion people and growing at a rate of 0.089 billion people per year.

65. The ripple's area and radius are related by $A(t) = \pi[r(t)]^2$. Taking derivatives and using the chain rule gives

$$\frac{dA}{dt} = \pi \cdot 2r \frac{dr}{dt}.$$

We know that $dr/dt = 10$ cm/sec, so when $r = 20$ cm we have

$$\frac{dA}{dt} = \pi \cdot 2 \cdot 20 \cdot 10 \text{cm}^2/\text{sec} = 400\pi \text{cm}^2/\text{sec}.$$

69. Recall that $v = dx/dt$. We want to find the acceleration, dv/dt, when $x = 2$. Differentiating the expression for v with respect to t using the chain rule and substituting for v gives

$$\frac{dv}{dt} = \frac{d}{dx}(x^2 + 3x - 2) \cdot \frac{dx}{dt} = (2x + 3)v = (2x + 3)(x^2 + 3x - 2).$$

Substituting $x = 2$ gives

$$\text{Acceleration} = \frac{dv}{dt}\bigg|_{x=2} = (2(2) + 3)(2^2 + 3 \cdot 2 - 2) = 56 \text{ cm/sec}^2.$$

Solutions for Section 3.5

Exercises

1.

Table 3.1

x	$\cos x$	Difference Quotient	$-\sin x$
0	1.0	-0.0005	0.0
0.1	0.995	-0.10033	-0.099833
0.2	0.98007	-0.19916	-0.19867
0.3	0.95534	-0.296	-0.29552
0.4	0.92106	-0.38988	-0.38942
0.5	0.87758	-0.47986	-0.47943
0.6	0.82534	-0.56506	-0.56464

5. $f'(x) = \cos(3x) \cdot 3 = 3\cos(3x)$.

9. $f'(x) = (2x)(\cos x) + x^2(-\sin x) = 2x\cos x - x^2\sin x$.

13. $z' = e^{\cos\theta} - \theta(\sin\theta)e^{\cos\theta}$.

17.

$$f(x) = (1 - \cos x)^{\frac{1}{2}}$$

$$f'(x) = \frac{1}{2}(1 - \cos x)^{-\frac{1}{2}}(-(-\sin x))$$

$$= \frac{\sin x}{2\sqrt{1 - \cos x}}.$$

21. $f'(x) = 2 \cdot [\sin(3x)] + 2x[\cos(3x)] \cdot 3 = 2\sin(3x) + 6x\cos(3x)$

25. $y' = 5\sin^4\theta\cos\theta$.

29. $h'(t) = 1 \cdot (\cos t) + t(-\sin t) + \frac{1}{\cos^2 t} = \cos t - t\sin t + \frac{1}{\cos^2 t}$.

33. Using the power and quotient rules gives

$$f'(x) = \frac{1}{2}\left(\frac{1 - \sin x}{1 - \cos x}\right)^{-1/2}\left[\frac{-\cos x(1 - \cos x) - (1 - \sin x)\sin x}{(1 - \cos x)^2}\right]$$

$$= \frac{1}{2}\sqrt{\frac{1 - \cos x}{1 - \sin x}}\left[\frac{-\cos x(1 - \cos x) - (1 - \sin x)\sin x}{(1 - \cos x)^2}\right]$$

$$= \frac{1}{2}\sqrt{\frac{1 - \cos x}{1 - \sin x}}\left[\frac{1 - \cos x - \sin x}{(1 - \cos x)^2}\right].$$

37. $w' = (\ln 2)(2^{2\sin x + e^x})(2\cos x + e^x)$.

41. $f'(w) = -2\cos w\sin w - \sin(w^2)(2w) = -2(\cos w\sin w + w\sin(w^2))$

Problems

45. We begin by taking the derivative of $y = \sin(x^4)$ and evaluating at $x = 10$:

$$\frac{dy}{dx} = \cos(x^4) \cdot 4x^3.$$

Evaluating $\cos(10,000)$ on a calculator (in radians) we see $\cos(10,000) < 0$, so we know that $dy/dx < 0$, and therefore the function is decreasing.

Next, we take the second derivative and evaluate it at $x = 10$;

$$\frac{d^2y}{dx^2} = \underbrace{\cos(x^4) \cdot (12x^2)}_{\text{negative}} + \underbrace{4x^3 \cdot (-\sin(x^4))(4x^3)}_{\substack{\text{positive, but much} \\ \text{larger in magnitude}}}.$$

From this we can see that $d^2y/dx^2 > 0$, thus the graph is concave up.

49. (a) When $\sqrt{\frac{k}{m}}t = \frac{\pi}{2}$ the spring is farthest from the equilibrium position. This occurs at time $t = \frac{\pi}{2}\sqrt{\frac{m}{k}}$

$v = A\sqrt{\frac{k}{m}}\cos\left(\sqrt{\frac{k}{m}}t\right)$, so the maximum velocity occurs when $t = 0$

$a = -A\frac{k}{m}\sin\left(\sqrt{\frac{k}{m}}t\right)$, so the maximum acceleration occurs when $\sqrt{\frac{k}{m}}t = \frac{3\pi}{2}$, which is at time $t = \frac{3\pi}{2}\sqrt{\frac{m}{k}}$

(b) $T = \frac{2\pi}{\sqrt{k/m}} = 2\pi\sqrt{\frac{m}{k}}$

(c) $\frac{dT}{dm} = \frac{2\pi}{\sqrt{k}} \cdot \frac{1}{2}m^{-\frac{1}{2}} = \frac{\pi}{\sqrt{km}}$

Since $\frac{dT}{dm} > 0$, an increase in the mass causes the period to increase.

53. (a) If $f(x) = \sin x$, then

$$f'(x) = \lim_{h \to 0} \frac{\sin(x+h) - \sin x}{h}$$
$$= \lim_{h \to 0} \frac{(\sin x \cos h + \sin h \cos x) - \sin x}{h}$$
$$= \lim_{h \to 0} \frac{\sin x(\cos h - 1) + \sin h \cos x}{h}$$
$$= \sin x \lim_{h \to 0} \frac{\cos h - 1}{h} + \cos x \lim_{h \to 0} \frac{\sin h}{h}.$$

(b) $\frac{\cos h - 1}{h} \to 0$ and $\frac{\sin h}{h} \to 1$, as $h \to 0$. Thus, $f'(x) = \sin x \cdot 0 + \cos x \cdot 1 = \cos x$.

(c) Similarly,

$$g'(x) = \lim_{h \to 0} \frac{\cos(x+h) - \cos x}{h}$$
$$= \lim_{h \to 0} \frac{(\cos x \cos h - \sin x \sin h) - \cos x}{h}$$
$$= \lim_{h \to 0} \frac{\cos x(\cos h - 1) - \sin x \sin h}{h}$$
$$= \cos x \lim_{h \to 0} \frac{\cos h - 1}{h} - \sin x \lim_{h \to 0} \frac{\sin h}{h}$$
$$= -\sin x.$$

Solutions for Section 3.6

Exercises

1. $f'(t) = \frac{2t}{t^2+1}$.

5. $f'(z) = -1(\ln z)^{-2} \cdot \frac{1}{z} = \frac{-1}{z(\ln z)^2}$.

9. $f'(x) = \frac{1}{e^x+1} \cdot e^x$.

13. $f'(x) = \frac{1}{e^{7x}} \cdot (e^{7x})7 = 7$.
   (Note also that $\ln(e^{7x}) = 7x$ implies $f'(x) = 7$.)

17. $f'(y) = \frac{2y}{\sqrt{1-y^4}}$.

21. $g'(t) = \frac{-\sin(\ln t)}{t}$.

25. Using the chain rule gives $r'(t) = \frac{2}{\sqrt{1-4t^2}}$.

29. Using the quotient rule gives

$$f'(x) = \frac{1 + \ln x - x(\frac{1}{x})}{(1 + \ln x)^2}$$
$$= \frac{\ln x}{(1 + \ln x)^2}.$$

33. Using the chain rule gives

$$T'(u) = \left[\frac{1}{1 + \left(\frac{u}{1+u}\right)^2}\right]\left[\frac{(1+u) - u}{(1+u)^2}\right]$$

$$= \frac{(1+u)^2}{(1+u)^2 + u^2}\left[\frac{1}{(1+u)^2}\right]$$

$$= \frac{1}{1 + 2u + 2u^2}.$$

Problems

37. Let

$$g(x) = \arcsin x$$

so

$$\sin[g(x)] = x.$$

Differentiating,

$$\cos[g(x)] \cdot g'(x) = 1$$

$$g'(x) = \frac{1}{\cos[g(x)]}$$

Using the fact that $\sin^2\theta + \cos^2\theta = 1$, and $\cos[g(x)] \geq 0$, since $-\frac{\pi}{2} \leq g(x) \leq \frac{\pi}{2}$, we get

$$\cos[g(x)] = \sqrt{1 - (\sin[g(x)])^2}.$$

Therefore,

$$g'(x) = \frac{1}{\sqrt{1 - (\sin[g(x)])^2}}.$$

Since $\sin[g(x)] = x$, we have

$$g'(x) = \frac{1}{\sqrt{1 - x^2}}, -1 < x < 1.$$

41. $\text{pH} = 2 = -\log x$ means $\log x = -2$ so $x = 10^{-2}$. Rate of change of pH with hydrogen ion concentration is

$$\frac{d}{dx}\text{pH} = -\frac{d}{dx}(\log x) = \frac{-1}{x(\ln 10)} = -\frac{1}{(10^{-2})\ln 10} = -43.4$$

45. (a) Let $g(x) = ax^2 + bx + c$ be our quadratic and $f(x) = \ln x$. For the best approximation, we want to find a quadratic with the same value as $\ln x$ at $x = 1$ and the same first and second derivatives as $\ln x$ at $x = 1$. $g'(x) = 2ax + b, g''(x) = 2a, f'(x) = \frac{1}{x}, f''(x) = -\frac{1}{x^2}$.

$$g(1) = a(1)^2 + b(1) + c \quad f(1) = 0$$
$$g'(1) = 2a(1) + b \quad f'(1) = 1$$
$$g''(1) = 2a \quad f''(1) = -1$$

Thus, we obtain the equations

$$a + b + c = 0$$
$$2a + b = 1$$
$$2a = -1$$

We find $a = -\frac{1}{2}, b = 2$ and $c = -\frac{3}{2}$. Thus our approximation is:

$$g(x) = -\frac{1}{2}x^2 + 2x - \frac{3}{2}$$

(b) From the graph below, we notice that around $x = 1$, the value of $f(x) = \ln x$ and the value of $g(x) = -\frac{1}{2}x^2 + 2x - \frac{3}{2}$ are very close.

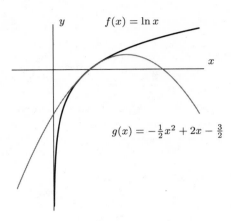

(c) $g(1.1) = 0.095$ $g(2) = 0.5$
Compare with $f(1.1) = 0.0953$, $f(2) = 0.693$.

49. If V is the volume of the balloon and r is its radius, then

$$V = \frac{4}{3}\pi r^3.$$

We want to know the rate at which air is being blown into the balloon, which is the rate at which the volume is increasing, dV/dt. We are told that

$$\frac{dr}{dt} = 2 \text{ cm/sec} \quad \text{when} \quad r = 10 \text{ cm}.$$

Using the chain rule, we have

$$\frac{dV}{dt} = \frac{dV}{dr} \cdot \frac{dr}{dt} = 4\pi r^2 \frac{dr}{dt}.$$

Substituting gives

$$\frac{dV}{dt} = 4\pi(10)^2 2 = 800\pi = 2513.3 \text{ cm}^3/\text{sec}.$$

53. (a) Using Pythagoras' theorem, we see

$$z^2 = 0.5^2 + x^2$$

so

$$z = \sqrt{0.25 + x^2}.$$

(b) We want to calculate dz/dt. Using the chain rule, we have

$$\frac{dz}{dt} = \frac{dz}{dx} \cdot \frac{dx}{dt} = \frac{2x}{2\sqrt{0.25 + x^2}} \frac{dx}{dt}.$$

Because the train is moving at 0.8 km/hr, we know that

$$\frac{dx}{dt} = 0.8 \text{ km/hr}.$$

At the moment we are interested in $z = 1$ km so

$$1^2 = 0.25 + x^2$$

giving

$$x = \sqrt{0.75} = 0.866 \text{ km}.$$

Therefore

$$\frac{dz}{dt} = \frac{2(0.866)}{2\sqrt{0.25 + 0.75}} \cdot 0.8 = 0.866 \cdot 0.8 = 0.693 \text{ km/min}.$$

(c) We want to know $d\theta/dt$, where θ is as shown in Figure 3.2. Since

$$\frac{x}{0.5} = \tan\theta$$

we know

$$\theta = \arctan\left(\frac{x}{0.5}\right),$$

so

$$\frac{d\theta}{dt} = \frac{1}{1 + (x/0.5)^2} \cdot \frac{1}{0.5}\frac{dx}{dt}.$$

We know that $dx/dt = 0.8$ km/min and, at the moment we are interested in, $x = \sqrt{0.75}$. Substituting gives

$$\frac{d\theta}{dt} = \frac{1}{1 + 0.75/0.25} \cdot \frac{1}{0.5} \cdot 0.8 = 0.4\,\text{radians/min.}$$

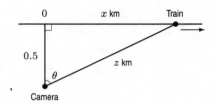

Figure 3.2

Solutions for Section 3.7

Exercises

1. We differentiate implicitly both sides of the equation with respect to x.

$$2x + 2y\frac{dy}{dx} = 0,$$

$$\frac{dy}{dx} = -\frac{2x}{2y} = -\frac{x}{y}.$$

5. We differentiate implicity with respect to x.

$$y + x\frac{dy}{dx} - 1 - \frac{3dy}{dx} = 0$$

$$(x - 3)\frac{dy}{dx} = 1 - y$$

$$\frac{dy}{dx} = \frac{1 - y}{x - 3}$$

9. We differentiate implicitly both sides of the equation with respect to x.

$$e^{x^2} + \ln y = 0$$

$$2xe^{x^2} + \frac{1}{y}\frac{dy}{dx} = 0$$

$$\frac{dy}{dx} = -2xye^{x^2}.$$

13. $\frac{2}{3}x^{-1/3} + \frac{2}{3}y^{-1/3} \cdot \frac{dy}{dx} = 0$, $\frac{dy}{dx} = -\frac{x^{-1/3}}{y^{-1/3}} = -\frac{y^{1/3}}{x^{1/3}}$.

17. Differentiating $\sin(xy) = x$ with respect to x gives

$$(y + xy')\cos(xy) = 1$$

or

$$xy'\cos(xy) = 1 - y\cos(xy)$$

so that

$$y' = \frac{1 - y\cos(xy)}{x\cos(xy)}.$$

As we move along the curve to the point $(1, \frac{\pi}{2})$, the value of $dy/dx \to \infty$, which tells us the tangent to the curve at $(1, \frac{\pi}{2})$ has infinite slope; the tangent is the vertical line $x = 1$.

21. First we must find the slope of the tangent, $\frac{dy}{dx}$, at $(1, e^2)$. Differentiating implicitly, we have:

$$\frac{1}{xy}\left(x\frac{dy}{dx} + y\right) = 2$$

$$\frac{dy}{dx} = \frac{2xy - y}{x}.$$

Evaluating dy/dx at $(1, e^2)$ yields $(2(1)e^2 - e^2)/1 = e^2$. Using the point-slope formula for the equation of the line, we have:

$$y - e^2 = e^2(x - 1),$$

or

$$y = e^2 x.$$

Problems

25. (a) Taking derivatives implicitly, we get

$$\frac{2}{25}x + \frac{2}{9}y\frac{dy}{dx} = 0$$

$$\frac{dy}{dx} = \frac{-9x}{25y}$$

(b) The slope is not defined anywhere along the line $y = 0$. This ellipse intersects that line in two places, $(-5, 0)$ and $(5, 0)$. (These are, of course, the "ends" of the ellipse where the tangent is vertical.)

29. Let the point of intersection of the tangent line with the smaller circle be (x_1, y_1) and the point of intersection with the larger be (x_2, y_2). Let the tangent line be $y = mx + c$. Then at (x_1, y_1) and (x_2, y_2) the slopes of $x^2 + y^2 = 1$ and $y^2 + (x - 3)^2 = 4$ are also m. The slope of $x^2 + y^2 = 1$ is found by implicit differentiation: $2x + 2yy' = 0$ so $y' = -x/y$. Similarly, the slope of $y^2 + (x - 3)^2 = 4$ is $y' = -(x - 3)/y$. Thus,

$$m = \frac{y_2 - y_1}{x_2 - x_1} = -\frac{x_1}{y_1} = -\frac{(x_2 - 3)}{y_2},$$

where $y_1 = \sqrt{1 - x_1^2}$ and $y_2 = \sqrt{4 - (x_2 - 3)^2}$. The positive values for y_1 and y_2 follow from Figure 3.3 and from our choice of $m > 0$. We obtain

$$\frac{x_1}{\sqrt{1 - x_1^2}} = \frac{x_2 - 3}{\sqrt{4 - (x_2 - 3)^2}}$$

$$\frac{x_1^2}{1 - x_1^2} = \frac{(x_2 - 3)^2}{4 - (x_2 - 3)^2}$$

$$x_1^2[4 - (x_2 - 3)^2] = (1 - x_1^2)(x_2 - 3)^2$$

$$4x_1^2 - (x_1^2)(x_2 - 3)^2 = (x_2 - 3)^2 - x_1^2(x_2 - 3)^2$$

$$4x_1^2 = (x_2 - 3)^2$$

$$2|x_1| = |x_2 - 3|.$$

From the picture $x_1 < 0$ and $x_2 < 3$. This gives $x_2 = 2x_1 + 3$ and $y_2 = 2y_1$. From

$$\frac{y_2 - y_1}{x_2 - x_1} = -\frac{x_1}{y_1},$$

substituting $y_1 = \sqrt{1 - x_1^2}$, $y_2 = 2y_1$ and $x_2 = 2x_1 + 3$ gives

$$x_1 = -\frac{1}{3}.$$

From $x_2 = 2x_1 + 3$ we get $x_2 = 7/3$. In addition, $y_1 = \sqrt{1 - x_1^2}$ gives $y_1 = 2\sqrt{2}/3$, and finally $y_2 = 2y_1$ gives $y_2 = 4\sqrt{2}/3$.

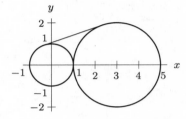

Figure 3.3

Solutions for Section 3.8

Exercises

1. Between times $t = 0$ and $t = 1$, x goes at a constant rate from 0 to 1 and y goes at a constant rate from 1 to 0. So the particle moves in a straight line from $(0, 1)$ to $(1, 0)$. Similarly, between times $t = 1$ and $t = 2$, it goes in a straight line to $(0, -1)$, then to $(-1, 0)$, then back to $(0, 1)$. So it traces out the diamond shown in Figure 3.4.

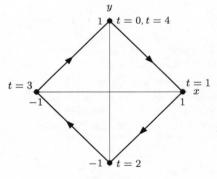

Figure 3.4

5. The particle moves clockwise: For $0 \leq t \leq \frac{\pi}{2}$, we have $x = \cos t$ decreasing and $y = -\sin t$ decreasing. Similarly, for the time intervals $\frac{\pi}{2} \leq t \leq \pi$, $\pi \leq t \leq \frac{3\pi}{2}$, and $\frac{3\pi}{2} \leq t \leq 2\pi$, we see that the particle moves clockwise.

9. Let $f(t) = \ln t$. Then $f'(t) = \frac{1}{t}$. The particle is moving counterclockwise when $f'(t) > 0$, that is, when $t > 0$. Any other time, when $t \leq 0$, the position is not defined.

13. One possible answer is $x = 2 + 5\cos t, y = 1 + 5\sin t, 0 \leq t \leq 2\pi$.

17. The parameterization $x = -3\cos t$, $y = 7\sin t$, $0 \leq t \leq 2\pi$, starts at the right point but sweeps out the ellipse in the wrong direction (the y-coordinate becomes positive as t increases). Thus, a possible parameterization is $x = -3\cos(-t) = -3\cos t, y = 7\sin(-t) = -7\sin t, 0 \leq t \leq 2\pi$.

21. We have $dx/dt = 2t$ and $dy/dt = 3t^2$. Therefore, the speed of the particle is

$$v = \sqrt{\left(\frac{dx}{dt}\right)^2 + \left(\frac{dy}{dt}\right)^2} = \sqrt{((2t)^2 + (3t^2)^2)} = |t| \cdot \sqrt{(4 + 9t^2)}.$$

The particle comes to a complete stop when its speed is 0, that is, if $t\sqrt{4 + 9t^2} = 0$, and so when $t = 0$.

25. At $t = 2$, the position is $(2^2, 2^3) = (4, 8)$, the velocity in the x-direction is $2 \cdot 2 = 4$, and the velocity in the y-direction is $3 \cdot 2^2 = 12$. So we want the line going through the point $(4, 8)$ at the time $t = 2$, with the given x- and y-velocities:

$$x = 4 + 4(t - 2), \quad y = 8 + 12(t - 2).$$

Problems

29. (a) The curve is a spiral as shown in Figure 3.5.

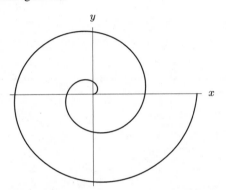

Figure 3.5: The spiral $x = t \cos t, y = t \sin t$
for $0 \leq t \leq 4\pi$

(b) At $t = 2$, the position is $(2 \cos 2, 2 \sin 2) = (-0.8323, 1.8186)$, and at $t = 2.01$ the position is $(2.01 \cos 2.01, 2.01 \sin 2.01) = (-0.8546, 1.8192)$. The distance between these points is

$$\sqrt{(-0.8546 - (-0.8323))^2 + (1.8192 - 1.8186)^2} \approx 0.022.$$

Thus the speed is approximately $0.022/0.01 \approx 2.2$.

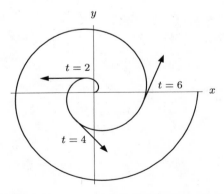

Figure 3.6: The spiral $x = t \cos t, y = t \sin t$ and three velocity vectors

(c) Evaluating the exact formula

$$v = \sqrt{(\cos t - t \sin t)^2 + (\sin t + t \cos t)^2}$$

gives :

$$v(2) = \sqrt{(-2.235)^2 + (0.077)^2} = 2.2363.$$

33. (a) C_1 has center at the origin and radius 5, so $a = b = 0, k = 5$ or -5.
   (b) C_2 has center at $(0, 5)$ and radius 5, so $a = 0, b = 5, k = 5$ or -5.
   (c) C_3 has center at $(10, -10)$, so $a = 10, b = -10$. The radius of C_3 is $\sqrt{10^2 + (-10)^2} = \sqrt{200}$, so $k = \sqrt{200}$ or $k = -\sqrt{200}$.

37. For $0 \le t \le 2\pi$

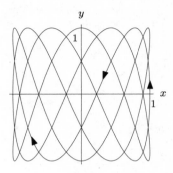

Solutions for Section 3.9

Exercises

1. With $f(x) = \sqrt{1 + x}$, the chain rule gives $f'(x) = 1/(2\sqrt{1 + x})$, so $f(0) = 1$ and $f'(0) = 1/2$. Therefore the tangent line approximation of f near $x = 0$,

$$f(x) \approx f(0) + f'(0)(x - 0),$$

becomes

$$\sqrt{1 + x} \approx 1 + \frac{x}{2}.$$

This means that, near $x = 0$, the function $\sqrt{1 + x}$ can be approximated by its tangent line $y = 1 + x/2$. (See Figure 3.7.)

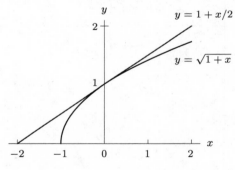

Figure 3.7

5. Let $f(x) = e^{-x}$. Then $f'(x) = -e^{-x}$. So $f(0) = 1$, $f'(0) = -e^0 = -1$. Therefore, $e^{-x} \approx f(0) + f'(0)x = 1 - x$.

Problems

9. (a) Let $f(x) = 1/(1 + x)$. Then $f'(x) = -1/(1 + x)^2$ by the chain rule. So $f(0) = 1$, and $f'(0) = -1$. Therefore, for x near 0, $1/(1 + x) \approx f(0) + f'(0)x = 1 - x$.
   (b) We know that for small y, $1/(1 + y) \approx 1 - y$. Let $y = x^2$; when x is small, so is $y = x^2$. Hence, for small x, $1/(1 + x^2) \approx 1 - x^2$.
   (c) Since the linearization of $1/(1 + x^2)$ is the line $y = 1$, and this line has a slope of 0, the derivative of $1/(1 + x^2)$ is zero at $x = 0$.

13. **(a)** Considering l as a constant, we have

$$T = f(g) = 2\pi\sqrt{\frac{l}{g}}.$$

Then,

$$f'(g) = 2\pi\sqrt{l}\left(-\frac{1}{2}g^{-3/2}\right) = -\pi\sqrt{\frac{l}{g^3}}.$$

Thus, local linearity gives

$$f(g + \Delta g) \approx f(g) - \pi\sqrt{\frac{l}{g^3}}(\Delta g).$$

Since $T = f(g)$ and $\Delta T = f(g + \Delta g) - f(g)$, we have

$$\Delta T \approx -\pi\sqrt{\frac{l}{g^3}}\Delta g = -2\pi\sqrt{\frac{l}{g}}\frac{\Delta g}{2g} = \frac{-T}{2}\frac{\Delta g}{g}.$$

$$\Delta T \approx \frac{-T}{2}\frac{\Delta g}{g}.$$

(b) If g increases by 1%, we know

$$\frac{\Delta g}{g} = 0.01.$$

Thus,

$$\frac{\Delta T}{T} \approx -\frac{1}{2}\frac{\Delta g}{g} = -\frac{1}{2}(0.01) = -0.005,$$

So, T decreases by 0.5%.

17. Note that

$$[f(g(x))]' = \lim_{h \to 0}\frac{f(g(x + h)) - f(g(x))}{h}.$$

Using the local linearizations of f and g, we get that

$$f(g(x + h)) - f(g(x)) \approx f\left(g(x) + g'(x)h\right) - f(g(x))$$
$$\approx f\left(g(x)\right) + f'(g(x))g'(x)h - f(g(x))$$
$$= f'(g(x))g'(x)h.$$

Therefore,

$$[f(g(x))]' = \lim_{h \to 0}\frac{f(g(x + h)) - f(g(x))}{h}$$
$$= \lim_{h \to 0}\frac{f'(g(x))g'(x)h}{h}$$
$$= \lim_{h \to 0}f'(g(x))g'(x) = f'(g(x))g'(x).$$

A more complete derivation can be given using the error term discussed in the section on Differentiability and Linear Approximation in Chapter 2. Adapting the notation of that section to this problem, we write

$$f(z + k) = f(z) + f'(z)k + E_f(k) \quad \text{and} \quad g(x + h) = g(x) + g'(x)h + E_g(h),$$

where $\displaystyle\lim_{h \to 0}\frac{E_g(h)}{h} = \lim_{k \to 0}\frac{E_f(k)}{k} = 0.$

Now we let $z = g(x)$ and $k = g(x + h) - g(x)$. Then we have $k = g'(x)h + E_g(h)$. Thus,

$$\frac{f(g(x + h)) - f(g(x))}{h} = \frac{f(z + k) - f(z)}{h}$$
$$= \frac{f(z) + f'(z)k + E_f(k) - f(z)}{h} = \frac{f'(z)k + E_f(k)}{h}$$
$$= \frac{f'(z)g'(x)h + f'(z)E_g(h)}{h} + \frac{E_f(k)}{k}\cdot\left(\frac{k}{h}\right)$$
$$= f'(z)g'(x) + \frac{f'(z)E_g(h)}{h} + \frac{E_f(k)}{k}\left[\frac{g'(x)h + E_g(h)}{h}\right]$$
$$= f'(z)g'(x) + \frac{f'(z)E_g(h)}{h} + \frac{g'(x)E_f(k)}{k} + \frac{E_g(h)\cdot E_f(k)}{h\cdot k}$$

Now, if $h \to 0$ then $k \to 0$ as well, and all the terms on the right except the first go to zero, leaving us with the term $f'(z)g'(x)$. Substituting $g(x)$ for z, we obtain

$$[f(g(x))]' = \lim_{h \to 0} \frac{f(g(x+h)) - f(g(x))}{h} = f'(g(x))g'(x).$$

Solutions for Section 3.10

Exercises

1. Since $f'(a) > 0$ and $g'(a) < 0$, l'Hopital's rule tells us that

$$\lim_{x \to a} \frac{f(x)}{g(x)} = \frac{f'(a)}{g'(a)} < 0.$$

5. The denominator approaches zero as x goes to zero and the numerator goes to zero even faster, so you should expect that the limit to be 0. You can check this by substituting several values of x close to zero. Alternatively, using l'Hopital's rule, we have

$$\lim_{x \to 0} \frac{x^2}{\sin x} = \lim_{x \to 0} \frac{2x}{\cos x} = 0.$$

9. The larger power dominates. Using l'Hopital's rule

$$\lim_{x \to \infty} \frac{x^5}{0.1x^7} = \lim_{x \to \infty} \frac{5x^4}{0.7x^6} = \lim_{x \to \infty} \frac{20x^3}{4.2x^5}$$
$$= \lim_{x \to \infty} \frac{60x^2}{21x^4} = \lim_{x \to \infty} \frac{120x}{84x^3} = \lim_{x \to \infty} \frac{120}{252x^2} = 0$$

so $0.1x^7$ dominates.

Problems

13. Let $f(x) = \ln x$ and $g(x) = 1/x$ so $f'(x) = 1/x$ and $g'(x) = -1/x^2$ and

$$\lim_{x \to 0+} \frac{\ln x}{1/x} = \lim_{x \to 0+} \frac{1/x}{-1/x^2} = \lim_{x \to 0+} \frac{x}{-1} = 0.$$

17. Let $f(x) = e^{-x}$ and $g(x) = \sin x$. Observe that as x increases, $f(x)$ approaches 0 but $g(x)$ oscillates between -1 and 1. Since $g(x)$ does not approach 0 in the limit, l'Hopital's rule does not apply. Because $g(x)$ is in the denominator and oscillates through 0 forever, the limit does not exist.

Solutions for Chapter 3 Review

Exercises

1. $f'(t) = \dfrac{d}{dt}\left(2te^t - \dfrac{1}{\sqrt{t}}\right) = 2e^t + 2te^t + \dfrac{1}{2t^{3/2}}.$

5. $g'(x) = \dfrac{d}{dx}\left(x^k + k^x\right) = kx^{k-1} + k^x \ln k.$

9. $s'(\theta) = \dfrac{d}{d\theta}\sin^2(3\theta - \pi) = 6\cos(3\theta - \pi)\sin(3\theta - \pi).$

13. $g'(x) = \dfrac{d}{dx}\left(x^{\frac{1}{2}} + x^{-1} + x^{-\frac{3}{2}}\right) = \dfrac{1}{2}x^{-\frac{1}{2}} - x^{-2} - \dfrac{3}{2}x^{-\frac{5}{2}}.$

17. Using the chain rule we get:

$$k'(\alpha) = e^{\tan(\sin \alpha)}(\tan(\sin \alpha))' = e^{\tan(\sin \alpha)} \cdot \frac{1}{\cos^2(\sin \alpha)} \cdot \cos \alpha.$$

21. $\dfrac{d}{dx}xe^{\tan x} = e^{\tan x} + xe^{\tan x}\dfrac{1}{\cos^2 x}.$

25. $\dfrac{dy}{dx} = (\ln 2)2^{\sin x}\cos x \cdot \cos x + 2^{\sin x}(-\sin x) = 2^{\sin x}\left((\ln 2)\cos^2 x - \sin x\right)$

29. Using the product rule and factoring gives $f'(t) = e^{-4kt}(\cos t - 4k\sin t)$.

33. Using the quotient rule gives

$$f'(s) = \frac{-2s\sqrt{a^2 + s^2} - \frac{s}{\sqrt{a^2+s^2}}(a^2 - s^2)}{(a^2 + s^2)}$$

$$= \frac{-2s(a^2 + s^2) - s(a^2 - s^2)}{(a^2 + s^2)^{3/2}}$$

$$= \frac{-2a^2 s - 2s^3 - a^2 s + s^3}{(a^2 + s^2)^{3/2}}$$

$$= \frac{-3a^2 s - s^3}{(a^2 + s^2)^{3/2}}.$$

37. Using the chain rule gives $r'(t) = \dfrac{\cos(\frac{t}{k})}{\sin(\frac{t}{k})}\left(\dfrac{1}{k}\right).$

41. Using the quotient and chain rules, we have

$$\frac{dy}{dx} = \frac{(ae^{ax} + ae^{-ax})(e^{ax} + e^{-ax}) - (e^{ax} - e^{-ax})(ae^{ax} - ae^{-ax})}{(e^{ax} + e^{-ax})^2}$$

$$= \frac{a(e^{ax} + e^{-ax})^2 - a(e^{ax} - e^{-ax})^2}{(e^{ax} + e^{-ax})^2}$$

$$= \frac{a[(e^{2ax} + 2 + e^{-2ax}) - (e^{2ax} - 2 + e^{-2ax})]}{(e^{ax} + e^{-ax})^2}$$

$$= \frac{4a}{(e^{ax} + e^{-ax})^2}$$

45. Since $\tan(\arctan(k\theta)) = k\theta$, because tan and arctan are inverse functions, we have $N'(\theta) = k$.

49. Since $\cos^2 y + \sin^2 y = 1$, we have $s(y) = \sqrt[3]{1 + 3} = \sqrt[3]{4}$. Thus $s'(y) = 0$.

53. We wish to find the slope $m = dy/dx$. To do this, we can implicitly differentiate the given formula in terms of x:

$$x^2 + 3y^2 = 7$$

$$2x + 6y\frac{dy}{dx} = \frac{d}{dx}(7) = 0$$

$$\frac{dy}{dx} = \frac{-2x}{6y} = \frac{-x}{3y}.$$

Thus, at $(2, -1)$, $m = -(2)/3(-1) = 2/3$.

57. Differentiating implicitly on both sides with respect to x,

$$a\cos(ay)\frac{dy}{dx} - b\sin(bx) = y + x\frac{dy}{dx}$$

$$(a\cos(ay) - x)\frac{dy}{dx} = y + b\sin(bx)$$

$$\frac{dy}{dx} = \frac{y + b\sin(bx)}{a\cos(ay) - x}.$$

Problems

61. (a) $H'(2) = r'(2)s(2) + r(2)s'(2) = -1 \cdot 1 + 4 \cdot 3 = 11.$

(b) $H'(2) = \dfrac{r'(2)}{2\sqrt{r(2)}} = \dfrac{-1}{2\sqrt{4}} = -\dfrac{1}{4}.$

(c) $H'(2) = r'(s(2))s'(2) = r'(1) \cdot 3$, but we don't know $r'(1)$.

(d) $H'(2) = s'(r(2))r'(2) = s'(4)r'(2) = -3.$

65. It makes sense to define the angle between two curves to be the angle between their tangent lines. (The tangent lines are the best linear approximations to the curves). See Figure 3.8. The functions $\sin x$ and $\cos x$ are equal at $x = \frac{\pi}{4}$.

$$\text{For } f_1(x) = \sin x, \quad f_1'\left(\frac{\pi}{4}\right) = \cos\left(\frac{\pi}{4}\right) = \frac{\sqrt{2}}{2}$$

$$\text{For } f_2(x) = \cos x, \quad f_2'\left(\frac{\pi}{4}\right) = -\sin\left(\frac{\pi}{4}\right) = -\frac{\sqrt{2}}{2}.$$

Using the point $\left(\frac{\pi}{4}, \frac{\sqrt{2}}{2}\right)$ for each tangent line we get $y = \frac{\sqrt{2}}{2}x + \frac{\sqrt{2}}{2}\left(1 - \frac{\pi}{4}\right)$ and $y = -\frac{\sqrt{2}}{2}x + \frac{\sqrt{2}}{2}\left(1 + \frac{\pi}{4}\right)$, respectively.

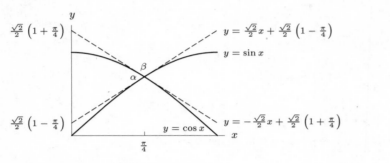

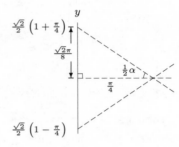

Figure 3.8   **Figure 3.9**

There are two possibilities of how to define the angle between the tangent lines, indicated by α and β above. The choice is arbitrary, so we will solve for both. To find the angle, α, we consider the triangle formed by these two lines and the y-axis. See Figure 3.9.

$$\tan\left(\frac{1}{2}\alpha\right) = \frac{\sqrt{2}\pi/8}{\pi/4} = \frac{\sqrt{2}}{2}$$

$$\frac{1}{2}\alpha = 0.61548 \text{ radians}$$

$$\alpha = 1.231 \text{ radians, or } 70.5^\circ.$$

Now let us solve for β, the other possible measure of the angle between the two tangent lines. Since α and β are supplementary, $\beta = \pi - 1.231 = 1.909$ radians, or 109.4°.

69. (a) $\dfrac{dg}{dr} = GM\dfrac{d}{dr}\left(\dfrac{1}{r^2}\right) = GM\dfrac{d}{dr}\left(r^{-2}\right) = GM(-2)r^{-3} = -\dfrac{2GM}{r^3}.$

(b) $\dfrac{dg}{dr}$ is the rate of change of acceleration due to the pull of gravity. The further away from the center of the earth, the weaker the pull of gravity is. So g is decreasing and therefore its derivative, $\dfrac{dg}{dr}$, is negative.

(c) By part (a),

$$\left.\frac{dg}{dr}\right|_{r=6400} = -\left.\frac{2GM}{r^3}\right|_{r=6400} = -\frac{2(6.67 \times 10^{-20})(6 \times 10^{24})}{(6400)^3} \approx -3.05 \times 10^{-6}.$$

(d) It is reasonable to assume that g is a constant near the surface of the earth.

73.

$$\frac{dy}{dt} = -7.5(0.507)\sin(0.507t) = -3.80\sin(0.507t)$$

(a) When $t = 6$, $\frac{dy}{dt} = -3.80\sin(0.507 \cdot 6) = -0.38$ meters/hour. So the tide is falling at 0.38 meters/hour.

(b) When $t = 9$, $\frac{dy}{dt} = -3.80\sin(0.507 \cdot 9) = 3.76$ meters/hour. So the tide is rising at 3.76 meters/hour.

(c) When $t = 12$, $\frac{dy}{dt} = -3.80\sin(0.507 \cdot 12) = 0.75$ meters/hour. So the tide is rising at 0.75 meters/hour.

(d) When $t = 18$, $\frac{dy}{dt} = -3.80\sin(0.507 \cdot 18) = -1.12$ meters/hour. So the tide is falling at 1.12 meters/hour.

77. Let r be the radius of the balloon. Then its volume, V, is

$$V = \frac{4}{3}\pi r^3.$$

We need to find the rate of change of V with respect to time, that is dV/dt. Since $V = V(r)$,

$$\frac{dV}{dr} = 4\pi r^2$$

so that by the chain rule,

$$\frac{dV}{dt} = \frac{dV}{dr}\frac{dr}{dt} = 4\pi r^2 \cdot 1.$$

When $r = 5$, $dV/dt = 100\pi$ cm^3/sec.

81. We want to find dP/dV. Solving $PV = k$ for P gives

$$P = k/V$$

so,

$$\frac{dP}{dV} = -\frac{k}{V^2}.$$

85. This problem can be solved by using either the quotient rule or the fact that

$$\frac{f'}{f} = \frac{d}{dx}(\ln f) \quad \text{and} \quad \frac{g'}{g} = \frac{d}{dx}(\ln g).$$

We use the second method. The relative rate of change of f/g is $(f/g)'/(f/g)$, so

$$\frac{(f/g)'}{f/g} = \frac{d}{dx}\ln\left(\frac{f}{g}\right) = \frac{d}{dx}(\ln f - \ln g) = \frac{d}{dx}(\ln f) - \frac{d}{dx}(\ln g) = \frac{f'}{f} - \frac{g'}{g}.$$

Thus, the relative rate of change of f/g is the difference between the relative rates of change of f and of g.

CAS Challenge Problems

89. **(a)** A CAS gives $h'(t) = 0$

(b) By the chain rule

$$h'(t) = \frac{\frac{d}{dt}\left(1 - \frac{1}{t}\right)}{1 - \frac{1}{t}} + \frac{\frac{d}{dt}\left(\frac{t}{t-1}\right)}{\frac{t}{t-1}} = \frac{\frac{1}{t^2}}{\frac{t-1}{t}} + \frac{\frac{1}{t-1} - \frac{t}{(t-1)^2}}{\frac{t}{t-1}}$$

$$= \frac{1}{t^2 - t} + \frac{(t-1) - t}{t^2 - t} = \frac{1}{t^2 - t} + \frac{-1}{t^2 - t} = 0.$$

(c) The expression inside the first logarithm is $1 - (1/t) = (t - 1)/t$. Using the property $\log A + \log B = \log(AB)$, we get

$$\ln\left(1 - \frac{1}{t}\right) + \ln\left(\frac{t}{t-1}\right) = \ln\left(\frac{t-1}{t}\right) + \ln\left(\frac{t}{t-1}\right)$$

$$= \ln\left(\frac{t-1}{t} \cdot \frac{t}{1-t}\right) = \ln 1 = 0.$$

Thus $h(t) = 0$, so $h'(t) = 0$ also.

CHECK YOUR UNDERSTANDING

1. True. Since $d(x^n)/dx = nx^{n-1}$, the derivative of a power function is a power function, so the derivative of a polynomial is a polynomial.

5. True. Since $f'(x)$ is the limit

$$f'(x) = \lim_{h \to 0}\frac{f(x+h) - f(x)}{h},$$

the function f must be defined for all x.

9. True; differentiating the equation with respect to x, we get

$$2y\frac{dy}{dx} + y + x\frac{dy}{dx} = 0.$$

Solving for dy/dx, we get that

$$\frac{dy}{dx} = \frac{-y}{2y + x}.$$

Thus dy/dx exists where $2y + x \neq 0$. Now if $2y + x = 0$, then $x = -2y$. Substituting for x in the original equation, $y^2 + xy - 1 = 0$, we get

$$y^2 - 2y^2 - 1 = 0.$$

This simplifies to $y^2 + 1 = 0$, which has no solutions. Thus dy/dx exists everywhere.

13. False; the fourth derivative of $\cos t + C$, where C is any constant, is indeed $\cos t$. But any function of the form $\cos t + p(t)$, where $p(t)$ is a polynomial of degree less than or equal to 3, also has its fourth derivative equal to $\cos t$. So $\cos t + t^2$ will work.

17. False; for example, if both f and g are constant functions, then the derivative of $f(g(x))$ is zero, as is the derivative of $f(x)$. Another example is $f(x) = 5x + 7$ and $g(x) = x + 2$.

21. False. Let $f(x) = e^{-x}$ and $g(x) = x^2$. Let $h(x) = f(g(x)) = e^{-x^2}$. Then $h'(x) = -2xe^{-x^2}$ and $h''(x) = (-2 + 4x^2)e^{-x^2}$. Since $h''(0) < 0$, clearly h is not concave up for all x.

CHAPTER FOUR

Solutions for Section 4.1

Exercises

1. We sketch a graph which is horizontal at the two critical points. One possibility is shown in Figure 4.1.

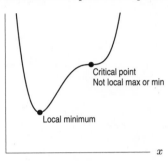

Figure 4.1

5. (a) A graph of $f(x) = e^{-x^2}$ is shown in Figure 4.2. It appears to have one critical point, at $x = 0$, and two inflection points, one between 0 and 1 and the other between 0 and -1.

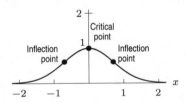

Figure 4.2

(b) To find the critical points, we set $f'(x) = 0$. Since $f'(x) = -2xe^{-x^2} = 0$, there is one solution, $x = 0$. The only critical point is at $x = 0$.

To find the inflection points, we first use the product rule to find $f''(x)$. We have

$$f''(x) = (-2x)(e^{-x^2}(-2x)) + (-2)(e^{-x^2}) = 4x^2e^{-x^2} - 2e^{-x^2}.$$

We set $f''(x) = 0$ and solve for x by factoring:

$$4x^2e^{-x^2} - 2e^{-x^2} = 0$$
$$(4x^2 - 2)e^{-x^2} = 0.$$

Since e^{-x^2} is never zero, we have

$$4x^2 - 2 = 0$$
$$x^2 = \frac{1}{2}$$
$$x = \pm 1/\sqrt{2}.$$

There are exactly two inflection points, at $x = 1/\sqrt{2} \approx 0.707$ and $x = -1/\sqrt{2} \approx -0.707$.

9. (a) A critical point occurs when $f'(x) = 0$. Since $f'(x)$ changes sign between $x = 2$ and $x = 3$, between $x = 6$ and $x = 7$, and between $x = 9$ and $x = 10$, we expect critical points at around $x = 2.5$, $x = 6.5$, and $x = 9.5$.

(b) Since $f'(x)$ goes from positive to negative at $x \approx 2.5$, a local maximum should occur there. Similarly, $x \approx 6.5$ is a local minimum and $x \approx 9.5$ a local maximum.

13.

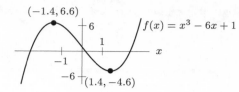

The graph of f above appears to be increasing for $x < -1.4$, decreasing for $-1.4 < x < 1.4$, and increasing for $x > 1.4$. There is a local maximum near $x = -1.4$ and local minimum near $x = 1.4$. The derivative of f is $f'(x) = 3x^2 - 6$. Thus $f'(x) = 0$ when $x^2 = 2$, that is $x = \pm\sqrt{2}$. This explains the critical points near $x = \pm 1.4$. Since $f'(x)$ changes from positive to negative at $x = -\sqrt{2}$, and from negative to positive at $x = \sqrt{2}$, there is a local maximum at $x = -\sqrt{2}$ and a local minimum at $x = \sqrt{2}$.

17.

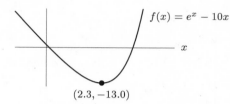

The graph of f above appears to be decreasing for $x < 2.3$ (almost like a straight line for $x < 0$), and increasing sharply for $x > 2.3$. Here $f'(x) = e^x - 10$, so $f'(x) = 0$ when $e^x = 10$, that is $x = \ln 10 = 2.302...$ This is the only place where $f'(x)$ changes sign, and it is a minimum of f. Notice that e^x is small for $x < 0$ so $f'(x) \approx -10$ for $x < 0$, which means the graph looks like a straight line of slope -10 for $x < 0$. However, e^x gets large quickly for $x > 0$, so $f'(x)$ gets large quickly for $x > \ln 10$, meaning the graph increases sharply there.

21. (13) The graph of $f(x) = x^3 - 6x + 1$ appears to be concave up for $x > 0$ and concave down for $x < 0$, with a point of inflection at $x = 0$. This is because $f''(x) = 6x$ is negative for $x < 0$ and positive for $x > 0$.

(14) Same answer as number 13.

(15) There appear to be three points of inflection at about $x = \pm 0.7$ and $x = 0$. This is because $f''(x) = 60x^3 - 30x = 30x(2x^2 - 1)$, which changes sign at $x = 0$ and $x = \pm 1/\sqrt{2}$.

(16) There appear to be points of inflection equally spaced about 3 units apart. This is because $f''(x) = -2\sin x$, which changes sign at $x = 0, \pm\pi, \pm 2\pi, \dots$.

(17) The graph appears to be concave up for all x. This is because $f''(x) = e^x > 0$ for all x.

(18) The graph appears to be concave up for all $x > 0$, and has almost periodic changes in concavity for $x < 0$. This is because for $x > 0$, $f''(x) = e^x - \sin x > 0$, and for $x < 0$, since e^x is small, $f''(x)$ changes sign at approximately the same values of x as $\sin x$.

(19) There appears to be a point of inflection for some $x < -0.71$, for $x = 0$, and for some $x > 0.71$. This is because $f'(x) = e^{-x^2}(1 - 2x^2)$ so

$$f''(x) = e^{-x^2}(-4x) + (1 - 2x^2)e^{-x^2}(-2x)$$
$$= e^{-x^2}(4x^3 - 6x).$$

Since $e^{-x^2} > 0$, this means $f''(x)$ has the same sign as $(4x^3 - 6x) = 2x(2x^2 - 3)$. Thus $f''(x)$ changes sign at $x = 0$ and $x = \pm\sqrt{3/2} \approx \pm 1.22$.

(20) The graph appears to be concave up for all x. This is because $f'(x) = 1 + \ln x$, so $f''(x) = 1/x$, which is greater than 0 for all $x > 0$.

Problems

25. We wish to have $f'(3) = 0$. Differentiating to find $f'(x)$ and then solving $f'(3) = 0$ for a gives:

$$f'(x) = x(ae^{ax}) + 1(e^{ax}) = e^{ax}(ax + 1)$$
$$f'(3) = e^{3a}(3a + 1) = 0$$
$$3a + 1 = 0$$
$$a = -\frac{1}{3}.$$

Thus, $f(x) = xe^{-x/3}$.

29. **(a)** Since the volume of water in the container is proportional to its depth, and the volume is increasing at a constant rate,

$$d(t) = \text{Depth at time } t = Kt,$$

where K is some positive constant. So the graph is linear, as shown in Figure 4.3. Since initially no water is in the container, we have $d(0) = 0$, and the graph starts from the origin.

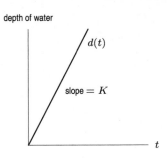

Figure 4.3

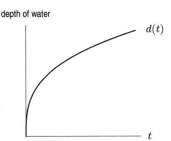

Figure 4.4

(b) As time increases, the additional volume needed to raise the water level by a fixed amount increases. Thus, although the depth, $d(t)$, of water in the cone at time t, continues to increase, it does so more and more slowly. This means $d'(t)$ is positive but decreasing, i.e., $d(t)$ is concave down. See Figure 4.4.

33. **(a)**   **(b)**

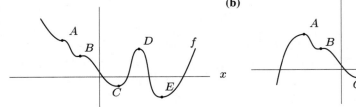

37. A has zeros where B has maxima and minima, so A could be a derivative of B. This is confirmed by comparing intervals on which B is increasing and A is positive. (They are the same.) So, C is either the derivative of A or the derivative of C is B. However, B does not have a zero at the point where C has a minimum, so B cannot be the derivative of C. Therefore, C is the derivative of A. So $B = f$, $A = f'$, and $C = f''$.

41.

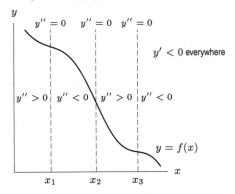

45. **(a)** Since $f''(x) > 0$ and $g''(x) > 0$ for all x, then $f''(x) + g''(x) > 0$ for all x, so $f(x) + g(x)$ is concave up for all x.
   (b) Nothing can be concluded about the concavity of $(f + g)(x)$. For example, if $f(x) = ax^2$ and $g(x) = bx^2$ with $a > 0$ and $b < 0$, then $(f + g)''(x) = a + b$. So $f + g$ is either always concave up, always concave down, or a straight line, depending on whether $a > |b|$, $a < |b|$, or $a = |b|$. More generally, it is even possible that $(f + g)(x)$ may have one or more changes in concavity.
   (c) It is possible to have infinitely many changes in concavity. Consider $f(x) = x^2 + \cos x$ and $g(x) = -x^2$. Since $f''(x) = 2 - \cos x$, we see that $f(x)$ is concave up for all x. Clearly $g(x)$ is concave down for all x. However, $f(x) + g(x) = \cos x$, which changes concavity an infinite number of times.

Solutions for Section 4.2

Exercises

1. We want a function of the form $y = a(x - h)^2 + k$, with $a < 0$ because the parabola opens downward. Since (h, k) is the vertex, we must take $h = 2$, $k = 5$, but we can take any negative value of a. Figure 4.5 shows the graph with $a = -1$, namely $y = -(x - 2)^2 + 5$.

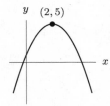

Figure 4.5: Graph of $y = -(x - 2)^2 + 5$

5. Since the vertical asymptote is $x = 2$, we have $b = -2$. The fact that the horizontal asymptote is $y = -5$ gives $a = -5$. So

$$y = \frac{-5x}{x - 2}.$$

9. Since the graph of the quartic polynomial is symmetric about the y-axis, the quartic must have only even powers and be of the form

$$y = ax^4 + bx^2 + c.$$

The y-intercept is 3, so $c = 3$. Differentiating gives

$$\frac{dy}{dx} = 4ax^3 + 2bx.$$

Since there is a maximum at $(1, 4)$, we have $dy/dx = 0$ if $x = 1$, so

$$4a(1)^3 + 2b(1) = 4a + 2b = 0 \qquad \text{so} \qquad b = -2a.$$

The fact that $dy/dx = 0$ if $x = -1$ gives us the same relationship

$$-4a - 2b = 0 \qquad \text{so} \qquad b = -2a.$$

We also know that $y = 4$ if $x = \pm 1$, so

$$a(1)^4 + b(1)^2 + 3 = a + b + 3 = 4 \qquad \text{so} \qquad a + b = 1.$$

Solving for a and b gives

$$a - 2a = 1 \qquad \text{so} \qquad a = -1 \text{ and } b = 2.$$

Finding d^2y/dx^2 so that we can check that $x = \pm 1$ are maxima, not minima, we see

$$\frac{d^2y}{dx^2} = 12ax^2 + 2b = -12x^2 + 4.$$

Thus $\dfrac{d^2y}{dx^2} = -8 < 0$ for $x = \pm 1$, so $x = \pm 1$ are maxima. See Figure 4.6.

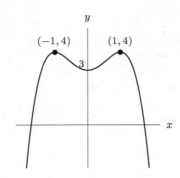

Figure 4.6: Graph of $y = -x^4 + 2x^2 + 3$

Problems

13. **(a)** We have $p'(x) = 3x^2 - a$, so

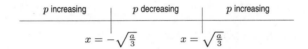

Local maximum: $p\left(-\sqrt{\frac{a}{3}}\right) = \frac{-a\sqrt{a}}{\sqrt{27}} + \frac{a\sqrt{a}}{\sqrt{3}} = +\frac{2a\sqrt{a}}{3\sqrt{3}}$

Local minimum: $p\left(\sqrt{\frac{a}{3}}\right) = -p\left(-\sqrt{\frac{a}{3}}\right) = -\frac{2a\sqrt{a}}{3\sqrt{3}}$

(b) Increasing the value of a moves the critical points of p away from the y-axis, and moves the critical values away from the x-axis. Thus, the "bumps" get further apart and higher. At the same time, increasing the value of a spreads the zeros of p further apart (while leaving the one at the origin fixed).

(c) See Figure 4.7

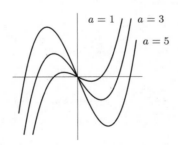

Figure 4.7

17. We begin by finding the intercepts, which occur where $f(x) = 0$, that is

$$x - k\sqrt{x} = 0$$
$$\sqrt{x}(\sqrt{x} - k) = 0$$

$$\text{so} \quad x = 0 \quad \text{or} \quad \sqrt{x} = k, \quad x = k^2.$$

So 0 and k^2 are the x-intercepts. Now we find the location of the critical points by setting $f'(x)$ equal to 0:

$$f'(x) = 1 - k\left(\frac{1}{2}x^{-(1/2)}\right) = 1 - \frac{k}{2\sqrt{x}} = 0.$$

This means

$$1 = \frac{k}{2\sqrt{x}}, \quad \text{so} \quad \sqrt{x} = \frac{1}{2}k, \quad \text{and} \quad x = \frac{1}{4}k^2.$$

We can use the second derivative to verify that $x = \frac{k^2}{4}$ is a local minimum. $f''(x) = 1 + \frac{k}{4x^{3/2}}$ is positive for all $x > 0$. So the critical point, $x = \frac{1}{4}k^2$, is 1/4 of the way between the x-intercepts, $x = 0$ and $x = k^2$. Since $f''(x) = \frac{1}{4}kx^{-3/2}$, $f''(\frac{1}{4}k^2) = 2/k^2 > 0$, this critical point is a minimum.

21. Since $f'(x) = abe^{-bx}$, we have $f'(x) > 0$ for all x. Therefore, f is increasing for all x. Since $f''(x) = -ab^2e^{-bx}$, we have $f''(x) < 0$ for all x. Therefore, f is concave down for all x.

25. (a) Figure 4.8 suggests that each graph decreases to a local minimum and then increases sharply. The local minimum appears to move to the right as k increases. It appears to move up until $k = 1$, and then to move back down.

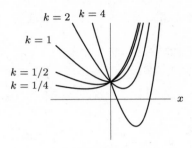

Figure 4.8

(b) $f'(x) = e^x - k = 0$ when $x = \ln k$. Since $f'(x) < 0$ for $x < \ln k$ and $f'(x) > 0$ for $x > \ln k$, f is decreasing to the left of $x = \ln k$ and increasing to the right, so f reaches a local minimum at $x = \ln k$.

(c) The minimum value of f is

$$f(\ln k) = e^{\ln k} - k(\ln k) = k - k\ln k.$$

Since we want to maximize the expression $k - k\ln k$, we can imagine a function $g(k) = k - k\ln k$. To maximize this function we simply take its derivative and find the critical points. Differentiating, we obtain

$$g'(k) = 1 - \ln k - k(1/k) = -\ln k.$$

Thus $g'(k) = 0$ when $k = 1$, $g'(k) > 0$ for $k < 1$, and $g'(k) < 0$ for $k > 1$. Thus $k = 1$ is a local maximum for $g(k)$. That is, the largest global minimum for f occurs when $k = 1$.

29.

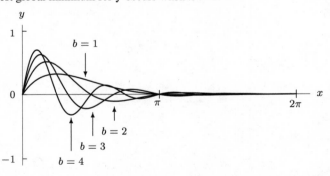

The larger the value of b, the narrower the humps and more humps per given region there are in the graph.

33. (a) The force is zero where

$$f(r) = -\frac{A}{r^2} + \frac{B}{r^3} = 0$$
$$Ar^3 = Br^2$$
$$r = \frac{B}{A}.$$

The vertical asymptote is $r = 0$ and the horizontal asymptote is the r-axis.

(b) To find critical points, we differentiate and set $f'(r) = 0$:

$$f'(r) = \frac{2A}{r^3} - \frac{3B}{r^4} = 0$$

$$2Ar^4 = 3Br^3$$

$$r = \frac{3B}{2A}.$$

Thus, $r = 3B/(2A)$ is the only critical point. Since $f'(r) < 0$ for $r < 3B/(2A)$ and $f'(r) > 0$ for $r > 3B/(2A)$, we see that $r = 3B/(2A)$ is a local minimum. At that point,

$$f\left(\frac{3B}{2A}\right) = -\frac{A}{9B^2/4A^2} + \frac{B}{27B^3/8A^3} = -\frac{4A^3}{27B^2}.$$

Differentiating again, we have

$$f''(r) = -\frac{6A}{r^4} + \frac{12B}{r^5} = -\frac{6}{r^5}(Ar - 2B).$$

So $f''(r) < 0$ where $r > 2B/A$ and $f''(r) > 0$ when $r < 2B/A$. Thus, $r = 2B/A$ is the only point of inflection. At that point

$$f\left(\frac{2B}{A}\right) = -\frac{A}{4B^2/A^2} + \frac{B}{8B^3/A^3} = -\frac{A^3}{8B^2}.$$

(c)

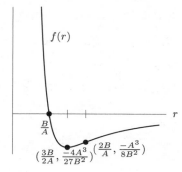

(d) (i) Increasing B means that the r-values of the zero, the minimum, and the inflection point increase, while the $f(r)$ values of the minimum and the point of inflection decrease in magnitude. See Figure 4.9.

(ii) Increasing A means that the r-values of the zero, the minimum, and the point of inflection decrease, while the $f(r)$ values of the minimum and the point of inflection increase in magnitude. See Figure 4.10.

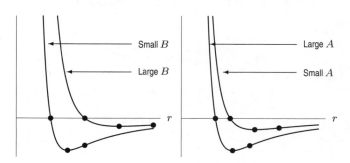

Figure 4.9: Increasing B **Figure 4.10**: Increasing A

Solutions for Section 4.3

Exercises

1.

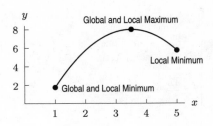

5. **(a)** Differentiating

$$f(x) = \sin^2 x - \cos x \quad \text{for } 0 \le x \le \pi$$
$$f'(x) = 2 \sin x \cos x + \sin x = (\sin x)(2 \cos x + 1)$$

$f'(x) = 0$ when $\sin x = 0$ or when $2 \cos x + 1 = 0$. Now, $\sin x = 0$ when $x = 0$ or when $x = \pi$. On the other hand, $2 \cos x + 1 = 0$ when $\cos x = -1/2$, which happens when $x = 2\pi/3$. So the critical points are $x = 0$, $x = 2\pi/3$, and $x = \pi$.

Note that $\sin x > 0$ for $0 < x < \pi$. Also, $2 \cos x + 1 < 0$ if $2\pi/3 < x \le \pi$ and $2 \cos x + 1 > 0$ if $0 < x < 2\pi/3$. Therefore,

$$f'(x) < 0 \quad \text{for} \quad \frac{2\pi}{3} < x < \pi$$
$$f'(x) > 0 \quad \text{for} \quad 0 < x < \frac{2\pi}{3}.$$

Thus f has a local maximum at $x = 2\pi/3$ and local minima at $x = 0$ and $x = \pi$.

(b) We have

$$f(0) = [\sin(0)]^2 - \cos(0) = -1$$
$$f\left(\frac{2\pi}{3}\right) = \left[\sin\left(\frac{2\pi}{3}\right)\right]^2 - \cos\frac{2\pi}{3} = 1.25$$
$$f(\pi) = [\sin(\pi)]^2 - \cos(\pi) = 1.$$

Thus the global maximum is at $x = 2\pi/3$, and the global minimum is at $x = 0$.

Problems

9. We have that $v(r) = a(R - r)r^2 = aRr^2 - ar^3$, and $v'(r) = 2aRr - 3ar^2 = 2ar(R - \frac{3}{2}r)$, which is zero if $r = \frac{2}{3}R$, or if $r = 0$, and so $v(r)$ has critical points there.

$v''(r) = 2aR - 6ar$, and thus $v''(0) = 2aR > 0$, which by the second derivative test implies that v has a minimum at $r = 0$. $v''(\frac{2}{3}R) = 2aR - 4aR = -2aR < 0$, and so by the second derivative test v has a maximum at $r = \frac{2}{3}R$. In fact, this is a global max of $v(r)$ since $v(0) = 0$ and $v(R) = 0$ at the endpoints.

13. We set $f'(r) = 0$ to find the critical points:

$$\frac{2A}{r^3} - \frac{3B}{r^4} = 0$$
$$\frac{2Ar - 3B}{r^4} = 0$$
$$2Ar - 3B = 0$$
$$r = \frac{3B}{2A}.$$

The only critical point is at $r = 3B/(2A)$. If $r > 3B/(2A)$, we have $f' > 0$ and if $r < 3B/(2A)$, we have $f' < 0$. Thus, the force between the atoms is minimized at $r = 3B/(2A)$.

17. A graph of F against θ is shown below.

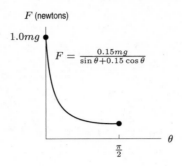

Taking the derivative:

$$\frac{dF}{d\theta} = -\frac{mg\mu(\cos\theta - \mu\sin\theta)}{(\sin\theta + \mu\cos\theta)^2}.$$

At a critical point, $dF/d\theta = 0$, so

$$\cos\theta - \mu\sin\theta = 0$$
$$\tan\theta = \frac{1}{\mu}$$
$$\theta = \arctan\left(\frac{1}{\mu}\right).$$

If $\mu = 0.15$, then $\theta = \arctan(1/0.15) = 1.422 \approx 81.5°$. To calculate the maximum and minimum values of F, we evaluate at this critical point and the endpoints:

$$\text{At } \theta = 0, \quad F = \frac{0.15mg}{\sin 0 + 0.15\cos 0} = 1.0mg \text{ newtons.}$$

$$\text{At } \theta = 1.422, \ F = \frac{0.15mg}{\sin(1.422) + 0.15\cos(1.422)} = 0.148mg \text{ newtons.}$$

$$\text{At } \theta = \pi/2, \quad F = \frac{0.15mg}{\sin(\frac{\pi}{2}) + 0.15\cos(\frac{\pi}{2})} = 0.15mg \text{ newtons.}$$

Thus, the maximum value of F is $1.0mg$ newtons when $\theta = 0$ (her arm is vertical) and the minimum value of F is $0.148mg$ newtons is when $\theta = 1.422$ (her arm is close to horizontal). See Figure 4.11.

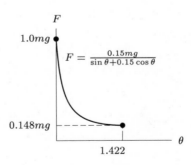

Figure 4.11

21. Let $y = e^{-x^2}$. Since $y' = -2xe^{-x^2}$, y is increasing for $x < 0$ and decreasing for $x > 0$. Hence $y = e^0 = 1$ is a global maximum.

When $x = \pm 0.3$, $y = e^{-0.09} \approx 0.9139$, which is a global minimum on the given interval. Thus $e^{-0.09} \le y \le 1$ for $|x| \le 0.3$.

25. The graph of $y = x + \sin x$ in Figure 4.12 suggests that the function is nondecreasing over the entire interval. You can confirm this by looking at the derivative:

$$y' = 1 + \cos x$$

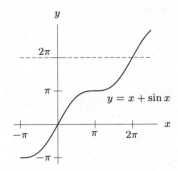

Figure 4.12: Graph of $y = x + \sin x$

Since $\cos x \geq -1$, we have $y' \geq 0$ everywhere, so y never decreases. This means that a lower bound for y is 0 (its value at the left endpoint of the interval) and an upper bound is 2π (its value at the right endpoint). That is, if $0 \leq x \leq 2\pi$:

$$0 \leq y \leq 2\pi.$$

These are the best bounds for y over the interval.

29. **(a)** At higher speeds, more energy is used so the graph rises to the right. The initial drop is explained by the fact that the energy it takes a bird to fly at very low speeds is greater than that needed to fly at a slightly higher speed. When it flies slightly faster, the amount of energy consumed decreases. But when it flies at very high speeds, the bird consumes a lot more energy (this is analogous to our swimming in a pool).

(b) $f(v)$ measures energy per second; $a(v)$ measures energy per meter. A bird traveling at rate v will in 1 second travel v meters, and thus will consume $v \cdot a(v)$ joules of energy in that 1 second period. Thus $v \cdot a(v)$ represents the energy consumption per second, and so $f(v) = v \cdot a(v)$.

(c) Since $v \cdot a(v) = f(v)$, $a(v) = f(v)/v$. But this ratio has the same value as the slope of a line passing from the origin through the point $(v, f(v))$ on the curve (see figure). Thus $a(v)$ is minimal when the slope of this line is minimal. To find the value of v minimizing $a(v)$, we solve $a'(v) = 0$. By the quotient rule,

$$a'(v) = \frac{vf'(v) - f(v)}{v^2}.$$

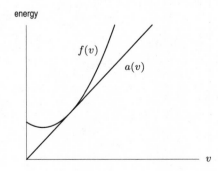

Thus $a'(v) = 0$ when $vf'(v) = f(v)$, or when $f'(v) = f(v)/v = a(v)$. Since $a(v)$ is represented by the slope of a line through the origin and a point on the curve, $a(v)$ is minimized when this line is tangent to $f(v)$, so that the slope $a(v)$ equals $f'(v)$.

(d) The bird should minimize $a(v)$ assuming it wants to go from one particular point to another, i.e. where the distance is set. Then minimizing $a(v)$ minimizes the total energy used for the flight.

33. Here is one possible graph of g:

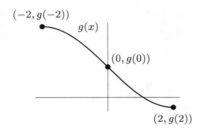

(a) From left to right, the graph of $g(x)$ starts "flat", decreases slowly at first then more rapidly, most rapidly at $x = 0$. The graph then continues to decrease but less and less rapidly until flat again at $x = 2$. The graph should exhibit symmetry about the point $(0, g(0))$.

(b) The graph has an inflection point at $(0, g(0))$ where the slope changes from negative and decreasing to negative and increasing.

(c) The function has a global maximum at $x = -2$ and a global minimum at $x = 2$.

(d) Since the function is decreasing over the interval $-2 \leq x \leq 2$

$$g(-2) = 5 > g(0) > g(2).$$

Since the function appears symmetric about $(0, g(0))$, we have

$$g(-2) - g(0) = g(0) - g(2).$$

Solutions for Section 4.4

Exercises

1. The fixed costs are $5000, the marginal cost per item is $2.40, and the price per item is $4.

5. Since fixed costs are represented by the vertical intercept, they are $1.1 million. The quantity that maximizes profit is about $q = 70$, and the profit achieved is $\$(3.7 - 2.5) = \$1.2$ million

Problems

9. (a) We know that Profit = Revenue − Cost, so differentiating with respect to q gives:

Marginal Profit = Marginal Revenue − Marginal Cost.

We see from the figure in the problem that just to the left of $q = a$, marginal revenue is less than marginal cost, so marginal profit is negative there. To the right of $q = a$ marginal revenue is greater than marginal cost, so marginal profit is positive there. At $q = a$ marginal profit changes from negative to positive. This means that profit is decreasing to the left of a and increasing to the right. The point $q = a$ corresponds to a local minimum of profit, and does not maximize profit. It would be a terrible idea for the company to set its production level at $q = a$.

(b) We see from the figure in the problem that just to the left of $q = b$ marginal revenue is greater than marginal cost, so marginal profit is positive there. Just to the right of $q = b$ marginal revenue is less than marginal cost, so marginal profit is negative there. At $q = b$ marginal profit changes from positive to negative. This means that profit is increasing to the left of b and decreasing to the right. The point $q = b$ corresponds to a local maximum of profit. In fact, since the area between the MC and MR curves in the figure in the text between $q = a$ and $q = b$ is bigger than the area between $q = 0$ and $q = a$, $q = b$ is in fact a global maximum.

13. (a) $N = 100 + 20x$, graphed in Figure 4.13.

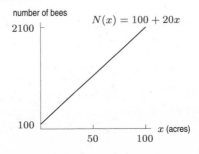

Figure 4.13

(b) $N'(x) = 20$ and its graph is just a horizontal line. This means that rate of increase of the number of bees with acres of clover is constant — each acre of clover brings 20 more bees.

On the other hand, $N(x)/x = 100/x + 20$ means that the average number of bees per acre of clover approaches 20 as more acres are put under clover. See Figure 4.14. As x increases, $100/x$ decreases to 0, so $N(x)/x$ approaches 20 (i.e. $N(x)/x \to 20$). Since the total number of bees is 20 per acre plus the original 100, the average number of bees per acre is 20 plus the 100 shared out over x acres. As x increases, the 100 are shared out over more acres, and so its contribution to the average becomes less. Thus the average number of bees per acre approaches 20 for large x.

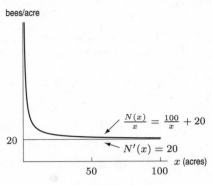

Figure 4.14

17. (a) Differentiating $C(q)$ gives

$$C'(q) = \frac{K}{a}q^{(1/a)-1}, \quad C''(q) = \frac{K}{a}\left(\frac{1}{a} - 1\right)q^{(1/a)-2}.$$

If $a > 1$, then $C''(q) < 0$, so C is concave down.

(b) We have

$$a(q) = \frac{C(q)}{q} = \frac{Kq^{1/a} + F}{q}$$

$$C'(q) = \frac{K}{a}q^{(1/a)-1}$$

so $a(q) = C'(q)$ means

$$\frac{Kq^{1/a} + F}{q} = \frac{K}{a}q^{(1/a)-1}.$$

Solving,

$$Kq^{1/a} + F = \frac{K}{a}q^{1/a}$$

$$K\left(\frac{1}{a} - 1\right)q^{1/a} = F$$

$$q = \left[\frac{Fa}{K(1-a)}\right]^a.$$

Solutions for Section 4.5

Exercises

1. We take the derivative, set it equal to 0, and solve for x:

$$\frac{dt}{dx} = \frac{1}{6} - \frac{1}{4} \cdot \frac{1}{2}\left((2000 - x)^2 + 600^2\right)^{-1/2} \cdot 2(2000 - x) = 0$$

$$(2000 - x) = \frac{2}{3}\left((2000 - x)^2 + 600^2\right)^{1/2}$$

$$(2000 - x)^2 = \frac{4}{9}\left((2000 - x)^2 + 600^2\right)$$

$$\frac{5}{9}(2000 - x)^2 = \frac{4}{9} \cdot 600^2$$

$$2000 - x = \sqrt{\frac{4}{5} \cdot 600^2} = \frac{1200}{\sqrt{5}}$$

$$x = 2000 - \frac{1200}{\sqrt{5}} \text{ feet.}$$

Note that $2000 - (1200/\sqrt{5}) \approx 1463$ feet, as given in the example.

Problems

5. (a) Suppose the height of the box is h. The box has six sides, four with area xh and two, the top and bottom, with area x^2. Thus,

$$4xh + 2x^2 = A.$$

So

$$h = \frac{A - 2x^2}{4x}.$$

Then, the volume, V, is given by

$$V = x^2h = x^2\left(\frac{A - 2x^2}{4x}\right) = \frac{x}{4}\left(A - 2x^2\right)$$

$$= \frac{A}{4}x - \frac{1}{2}x^3.$$

   (b) The graph is shown in Figure 4.15. We are assuming A is a positive constant. Also, we have drawn the whole graph, but we should only consider $V > 0$, $x > 0$ as V and x are lengths.

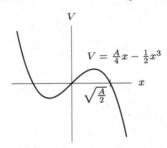

Figure 4.15

   (c) To find the maximum, we differentiate, regarding A as a constant:

$$\frac{dV}{dx} = \frac{A}{4} - \frac{3}{2}x^2.$$

So $dV/dx = 0$ if

$$\frac{A}{4} - \frac{3}{2}x^2 = 0$$

$$x = \pm\sqrt{\frac{A}{6}}.$$

For a real box, we must use $x = \sqrt{A/6}$. Figure 4.15 makes it clear that this value of x gives the maximum. Evaluating at $x = \sqrt{A/6}$, we get

$$V = \frac{A}{4}\sqrt{\frac{A}{6}} - \frac{1}{2}\left(\sqrt{\frac{A}{6}}\right)^3 = \frac{A}{4}\sqrt{\frac{A}{6}} - \frac{1}{2} \cdot \frac{A}{6}\sqrt{\frac{A}{6}} = \left(\frac{A}{6}\right)^{3/2}.$$

9. Figure 4.16 shows the the pool has dimensions x by y and the deck extends 5 feet at either side and 10 feet at the ends of the pool.

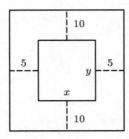

Figure 4.16

The dimensions of the plot of land containing the pool are then $(x + 5 + 5)$ by $(y + 10 + 10)$. The area of the land is then

$$A = (x + 10)(y + 20),$$

which is to be minimized. We also are told that the area of the pool is $xy = 1800$, so

$$y = 1800/x$$

and

$$A = (x + 10) \left(\frac{1800}{x} + 20 \right)$$
$$= 1800 + 20x + \frac{18000}{x} + 200.$$

We find dA/dx and set it to zero to get

$$\frac{dA}{dx} = 20 - \frac{18000}{x^2} = 0$$
$$20x^2 = 18000$$
$$x^2 = 900$$
$$x = 30 \text{ feet.}$$

Since $A \to \infty$ as $x \to 0^+$ and as $x \to \infty$, this critical point must be a global minimum. Also, $y = 1800/30 = 60$ feet. The plot of land is therefore $(30 + 10) = 40$ by $(60 + 20) = 80$ feet.

13. Any point on the curve can be written (x, x^2). The distance between such a point and $(3, 0)$ is given by

$$s(x) = \sqrt{(3 - x)^2 + (0 - x^2)^2} = \sqrt{(3 - x)^2 + x^4}.$$

Plotting this function in Figure 4.17, we see that there is a minimum near $x = 1$.

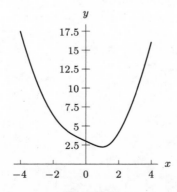

Figure 4.17

To find the value of x that minimizes the distance we can instead minimize the function $Q = s^2$ (the derivative is simpler). Then we have

$$Q(x) = (3-x)^2 + x^4.$$

Differentiating $Q(x)$ gives

$$\frac{dQ}{dx} = -6 + 2x + 4x^3.$$

Plotting the function $4x^3 + 2x - 6$ shows that there is one real solution at $x = 1$, which can be verified by substitution; the required coordinates are therefore $(1, 1)$. Because $Q''(x) = 2 + 12x^2$ is always positive, $x = 1$ is indeed the minimum. See Figure 4.18.

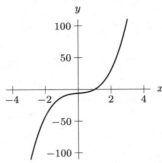

Figure 4.18

17. If v is the speed of the boat in miles per hour, then

$$\text{Cost of fuel per hour (in \$/hour)} = kv^3,$$

where k is the constant of proportionality. To find k, use the information that the boat uses \$100 worth of fuel per hour when cruising at 10 miles per hour: $100 = k10^3$, so $k = 100/10^3 = 0.1$. Thus,

$$\text{Cost of fuel per hour (in \$/hour)} = 0.1v^3.$$

From the given information, we also have

$$\text{Cost of other operations (labor, maintenance, etc.) per hour (in \$/hour)} = 675.$$

So

$$\text{Total Cost per hour (in \$/hour)} = \text{Cost of fuel (in \$/hour)} + \text{Cost of other (in \$/hour)}$$
$$= 0.1v^3 + 675.$$

However, we want to find the Cost per *mile*, which is the Total Cost per *hour* divided by the number of miles that the ferry travels in one hour. Since v is the speed in miles/hour at which the ferry travels, the number of miles that the ferry travels in one hour is simply v miles. Let $C = $ Cost per *mile*. Then

$$\text{Cost per \textit{mile} (in \$/mile)} = \frac{\text{Total Cost per \textit{hour} (in \$/hour)}}{\text{Distance traveled per hour (in miles/hour)}}$$

$$C = \frac{0.1v^3 + 675}{v} = 0.1v^2 + \frac{675}{v}.$$

We also know that $0 < v < \infty$. To find the speed at which Cost per *mile* is minimized, set

$$\frac{dC}{dv} = 2(0.1)v - \frac{675}{v^2} = 0$$

so

$$2(0.1)v = \frac{675}{v^2}$$

$$v^3 = \frac{675}{2(0.1)} = 3375$$

$$v = 15 \text{ miles/hour.}$$

Since

$$\frac{d^2C}{dv^2} = 0.2 + \frac{2(675)}{v^3} > 0$$

for $v > 0$, $v = 15$ gives a local minimum for C by the second-derivative test. Since this is the only critical point for $0 < v < \infty$, it must give a global minimum.

21. Let x be as indicated in the figure in the text. Then the distance from S to Town 1 is $\sqrt{1 + x^2}$ and the distance from S to Town 2 is $\sqrt{(4 - x)^2 + 4^2} = \sqrt{x^2 - 8x + 32}$.

$$\text{Total length of pipe} = f(x) = \sqrt{1 + x^2} + \sqrt{x^2 - 8x + 32}.$$

We want to look for critical points of f. The easiest way is to graph f and see that it has a local minimum at about $x = 0.8$ miles. Alternatively, we can use the formula:

$$\begin{aligned}
f'(x) &= \frac{2x}{2\sqrt{1 + x^2}} + \frac{2x - 8}{2\sqrt{x^2 - 8x + 32}} \\
&= \frac{x}{\sqrt{1 + x^2}} + \frac{x - 4}{\sqrt{x^2 - 8x + 32}} \\
&= \frac{x\sqrt{x^2 - 8x + 32} + (x - 4)\sqrt{1 + x^2}}{\sqrt{1 + x^2}\sqrt{x^2 - 8x + 32}} = 0.
\end{aligned}$$

$f'(x)$ is equal to zero when the numerator is equal to zero.

$$\begin{aligned}
x\sqrt{x^2 - 8x + 32} + (x - 4)\sqrt{1 + x^2} &= 0 \\
x\sqrt{x^2 - 8x + 32} &= (4 - x)\sqrt{1 + x^2}.
\end{aligned}$$

Squaring both sides and simplifying, we get

$$\begin{aligned}
x^2(x^2 - 8x + 32) &= (x^2 - 8x + 16)(14x^2) \\
x^4 - 8x^3 + 32x^2 &= x^4 - 8x^3 + 17x^2 - 8x + 16 \\
15x^2 + 8x - 16 &= 0, \\
(3x + 4)(5x - 4) &= 0.
\end{aligned}$$

So $x = 4/5$. (Discard $x = -4/3$ since we are only interested in x between 0 and 4, between the two towns.) Using the second derivative test, we can verify that $x = 4/5$ is a local minimum.

25. (a) Since $RB' = x$ and $A'R = c - x$, we have

$$AR = \sqrt{a^2 + (c - x)^2} \quad \text{and} \quad RB = \sqrt{b^2 + x^2}.$$

See Figure 4.19.

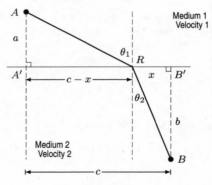

Figure 4.19

The time traveled, T, is given by

$$\begin{aligned}
T = \text{Time } AR + \text{Time } RB &= \frac{\text{Distance } AR}{v_1} + \frac{\text{Distance } RB}{v_2} \\
&= \frac{\sqrt{a^2 + (c - x)^2}}{v_1} + \frac{\sqrt{b^2 + x^2}}{v_2}.
\end{aligned}$$

(b) Let us calculate dT/dx:

$$\frac{dT}{dx} = \frac{-2(c-x)}{2v_1\sqrt{a^2+(c-x)^2}} + \frac{2x}{2v_2\sqrt{b^2+x^2}}.$$

At the minimum $dT/dx = 0$, so

$$\frac{c-x}{v_1\sqrt{a^2+(c-x)^2}} = \frac{x}{v_2\sqrt{b^2+x^2}}.$$

But we have

$$\sin\theta_1 = \frac{c-x}{\sqrt{a^2+(c-x)^2}} \quad\text{and}\quad \sin\theta_2 = \frac{x}{\sqrt{b^2+x^2}}.$$

Therefore, setting $dT/dx = 0$ tells us that

$$\frac{\sin\theta_1}{v_1} = \frac{\sin\theta_2}{v_2}$$

which gives

$$\frac{\sin\theta_1}{\sin\theta_2} = \frac{v_1}{v_2}.$$

Solutions for Section 4.6

Exercises

1. Using the chain rule, $\dfrac{d}{dx}\left(\cosh(2x)\right) = (\sinh(2x))\cdot 2 = 2\sinh(2x)$.

5. Using the chain rule,

$$\frac{d}{dt}\left(\cosh^2 t\right) = 2\cosh t \cdot \sinh t.$$

9. Substitute $x = 0$ into the formula for $\sinh x$. This yields

$$\sinh 0 = \frac{e^0 - e^{-0}}{2} = \frac{1-1}{2} = 0.$$

Problems

13. First we observe that

$$\sinh(2x) = \frac{e^{2x} - e^{-2x}}{2}.$$

Now let's calculate

$$(\sinh x)(\cosh x) = \left(\frac{e^x - e^{-x}}{2}\right)\left(\frac{e^x + e^{-x}}{2}\right)$$
$$= \frac{(e^x)^2 - (e^{-x})^2}{4}$$
$$= \frac{e^{2x} - e^{-2x}}{4}$$
$$= \frac{1}{2}\sinh(2x).$$

Thus, we see that

$$\sinh(2x) = 2\sinh x \cosh x.$$

17. (a) The graph in Figure 4.20 looks like the graph of $y = \cosh x$, with the minimum at about $(0.5, 6.3)$.

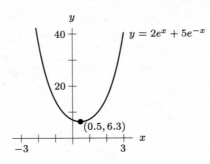

Figure 4.20

(b) We want to write

$$y = 2e^x + 5e^{-x} = A\cosh(x - c) = \frac{A}{2}e^{x-c} + \frac{A}{2}e^{-(x-c)}$$
$$= \frac{A}{2}e^x e^{-c} + \frac{A}{2}e^{-x}e^c$$
$$= \left(\frac{Ae^{-c}}{2}\right)e^x + \left(\frac{Ae^c}{2}\right)e^{-x}.$$

Thus, we need to choose A and c so that

$$\frac{Ae^{-c}}{2} = 2 \quad \text{and} \quad \frac{Ae^c}{2} = 5.$$

Dividing gives

$$\frac{Ae^c}{Ae^{-c}} = \frac{5}{2}$$
$$e^{2c} = 2.5$$
$$c = \frac{1}{2}\ln 2.5 \approx 0.458.$$

Solving for A gives

$$A = \frac{4}{e^{-c}} = 4e^c \approx 6.325.$$

Thus,

$$y = 6.325\cosh(x - 0.458).$$

Rewriting the function in this way shows that the graph in part (a) is the graph of $\cosh x$ shifted to the right by 0.458 and stretched vertically by a factor of 6.325.

21.

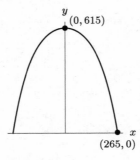

We know $x = 0$ and $y = 615$ at the top of the arch, so

$$615 = b - a\cosh(0/a) = b - a.$$

This means $b = a + 615$. We also know that $x = 265$ and $y = 0$ where the arch hits the ground, so

$$0 = b - a\cosh(265/a) = a + 615 - a\cosh(265/a).$$

We can solve this equation numerically on a calculator and get $a \approx 100$, which means $b \approx 715$. This results in the equation

$$y \approx 715 - 100\cosh\left(\frac{x}{100}\right).$$

Solutions for Section 4.7

Exercises

1. False. For example, if $f(x) = x^3$, then $f'(0) = 0$, so $x = 0$ is a critical point, but $x = 0$ is neither a local maximum nor a local minimum.

5. False. The horse that wins the race may have been moving faster for some, but not all, of the race. The Racetrack Principle guarantees the converse—that if the horses start at the same time and one moves faster throughout the race, then that horse wins.

9. No, it does not satisfy the hypotheses. The function does not appear to be differentiable. There appears to be no tangent line, and hence no derivative, at the "corner."

   No, it does not satisfy the conclusion as there is no horizontal tangent.

Problems

13. Let $f(x) = \sin x$ and $g(x) = x$. Then $f(0) = 0$ and $g(0) = 0$. Also $f'(x) = \cos x$ and $g'(x) = 1$, so for all $x \geq 0$ we have $f'(x) \leq g'(x)$. So the graphs of f and g both go through the origin and the graph of f climbs slower than the graph of g. Thus the graph of f is below the graph of g for $x \geq 0$ by the Racetrack Principle. In other words, $\sin x \leq x$ for $x \geq 0$.

17. The Decreasing Function Theorem is: Suppose that f is continuous on $[a, b]$ and differentiable on (a, b). If $f'(x) < 0$ on (a, b), then f is decreasing on $[a, b]$. If $f'(x) \leq 0$ on (a, b), then f is nonincreasing on $[a, b]$.

   To prove the theorem, we note that if f is decreasing then $-f$ is increasing and vice-versa. Similarly, if f is nonincreasing, then $-f$ is nondecreasing. Thus if $f'(x) < 0$, then $-f'(x) > 0$, so $-f$ is increasing, which means f is decreasing. And if $f'(x) \leq 0$, then $-f'(x) \geq 0$, so $-f$ is nondecreasing, which means f is nonincreasing.

21. By the Mean Value Theorem, Theorem 4.3, there is a number c, with $0 < c < 1$, such that

$$f'(c) = \frac{f(1) - f(0)}{1 - 0}.$$

   Since $f(1) - f(0) > 0$, we have $f'(c) > 0$.

   Alternatively if $f'(c) \leq 0$ for all c in $(0, 1)$, then by the Increasing Function Theorem, $f(0) \geq f(1)$.

25. If $f'(x) = 0$, then both $f'(x) \geq 0$ and $f'(x) \leq 0$. By the Increasing and Decreasing Function Theorems, f is both nondecreasing and nonincreasing, so f is constant.

29. **(a)** Since $f''(x) \geq 0$, $f'(x)$ is nondecreasing on (a, b). Thus $f'(c) \leq f'(x)$ for $c \leq x < b$ and $f'(x) \leq f'(c)$ for $a < x \leq c$.

   (b) Let $g(x) = f(c) + f'(c)(x - c)$ and $h(x) = f(x)$. Then $g(c) = f(c) = h(c)$, and $g'(x) = f'(c)$ and $h'(x) = f'(x)$. If $c \leq x < b$, then $g'(x) \leq h'(x)$, and if $a < x \leq c$, then $g'(x) \geq h'(x)$, by (a). By the Racetrack Principle, $g(x) \leq h'(x)$ for $c \leq x < b$ and for $a < x \leq c$, as we wanted.

Solutions for Chapter 4 Review

Exercises

1.

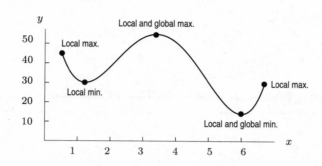

5. (a) Decreasing for $x < 0$, increasing for $0 < x < 4$, and decreasing for $x > 4$.
   (b) $f(0)$ is a local minimum, and $f(4)$ is a local maximum.

9. (a) First we find f' and f'':

$$f'(x) = -e^{-x} \sin x + e^{-x} \cos x$$
$$f''(x) = e^{-x} \sin x - e^{-x} \cos x$$
$$-e^{-x} \cos x - e^{-x} \sin x$$
$$= -2e^{-x} \cos x$$

   (b) The critical points are $x = \pi/4, 5\pi/4$, since $f'(x) = 0$ here.
   (c) The inflection points are $x = \pi/2, 3\pi/2$, since f'' changes sign at these points.
   (d) At the endpoints, $f(0) = 0$, $f(2\pi) = 0$. So we have $f(\pi/4) = (e^{-\pi/4})(\sqrt{2}/2)$ as the global maximum; $f(5\pi/4) = -e^{-5\pi/4}(\sqrt{2}/2)$ as the global minimum.
   (e)

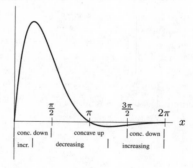

13. As $x \to -\infty$, $e^{-x} \to \infty$, so $xe^{-x} \to -\infty$. Thus $\lim_{x \to -\infty} xe^{-x} = -\infty$.
   As $x \to \infty$, $\frac{x}{e^x} \to 0$, since e^x grows much more quickly than x. Thus $\lim_{x \to \infty} xe^{-x} = 0$.
   Using the product rule,

$$f'(x) = e^{-x} - xe^{-x} = (1 - x)e^{-x},$$

which is zero when $x = 1$, negative when $x > 1$, and positive when $x < 1$. Thus $f(1) = 1/e^1 = 1/e$ is a local maximum.
   Again, using the product rule,

$$f''(x) = -e^{-x} - e^{-x} + xe^{-x}$$
$$= xe^{-x} - 2e^{-x}$$
$$= (x - 2)e^{-x},$$

which is zero when $x = 2$, positive when $x > 2$, and negative when $x < 2$, giving an inflection point at $(2, \frac{2}{e^2})$. With the above, we have the following diagram:

$y' > 0$ $y' < 0$

increasing   decreasing

$x = 1$

$y'' < 0$ $y'' > 0$

concave down   concave up

$x = 2$

The graph of f is shown below.

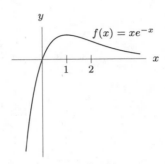

$f(x) = xe^{-x}$

and $f(x)$ has one global maximum at $1/e$ and no local or global minima.

17. $\lim\limits_{x \to +\infty} f(x) = +\infty$, $\lim\limits_{x \to -\infty} f(x) = 0$.

$y = 0$ is the horizontal asymptote.

$f'(x) = 2xe^{5x} + 5x^2 e^{5x} = xe^{5x}(5x + 2)$.

Thus, $x = -\frac{2}{5}$ and $x = 0$ are the critical points.

$$f''(x) = 2e^{5x} + 2xe^{5x} \cdot 5 + 10xe^{5x} + 25x^2 e^{5x}$$
$$= e^{5x}(25x^2 + 20x + 2).$$

So, $x = \dfrac{-2 \pm \sqrt{2}}{5}$ are inflection points.

x		$\frac{-2-\sqrt{2}}{5}$		$-\frac{2}{5}$		$\frac{-2+\sqrt{2}}{5}$		0	
f'	+	+	+	0	−	−	−	0	+
f''	+	0	−	−	−	0	+	+	+
f	↗⌣		↗⌢		↘⌢		↘⌣		↗⌣

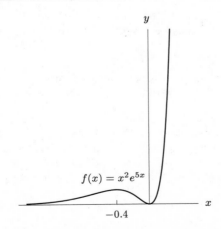

$f(x) = x^2 e^{5x}$

-0.4

So, $f(-\frac{2}{5})$ is a local maximum; $f(0)$ is a local and global minimum.

Problems

21. We want the maximum value of $r(t) = ate^{-bt}$ to be 0.3 ml/sec and to occur at $t = 0.5$ sec. Differentiating gives

$$r'(t) = ae^{-bt} - abte^{-bt},$$

so $r'(t) = 0$ when

$$ae^{-bt}(1 - bt) = 0 \qquad \text{or} \qquad t = \frac{1}{b}.$$

Since the maximum occurs at $t = 0.5$, we have

$$\frac{1}{b} = 0.5 \qquad \text{so} \qquad b = 2.$$

Thus, $r(t) = ate^{-2t}$. The maximum value of r is given by

$$r(0.5) = a(0.5)e^{-2(0.5)} = 0.5ae^{-1}.$$

Since the maximum value of r is 0.3, we have

$$0.5ae^{-1} = 0.3 \qquad \text{so} \qquad a = \frac{0.3e}{0.5} = 1.63.$$

Thus, $r(t) = 1.63te^{-2t}$ ml/sec.

25. The local maxima and minima of f correspond to places where f' is zero and changes sign or, possibly, to the endpoints of intervals in the domain of f. The points at which f changes concavity correspond to local maxima and minima of f'. The change of sign of f', from positive to negative corresponds to a maximum of f and change of sign of f' from negative to positive corresponds to a minimum of f.

29. Since the volume is fixed at 200 ml (i.e. 200 cm^3), we can solve the volume expression for h in terms of r to get (with h and r in centimeters)

$$h = \frac{200 \cdot 3}{7\pi r^2}.$$

Using this expression in the surface area formula we arrive at

$$S = 3\pi r \sqrt{r^2 + \left(\frac{600}{7\pi r^2}\right)^2}$$

By plotting $S(r)$ we see that there is a minimum value near $r = 2.7$ cm.

33. Let $f(x) = x \sin x$. Then $f'(x) = x \cos x + \sin x$.

$f'(x) = 0$ when $x = 0$, $x \approx 2$, and $x \approx 5$. The latter two estimates we can get from the graph of $f'(x)$.

Zooming in (or using some other approximation method), we can find the zeros of $f'(x)$ with more precision. They are (approximately) 0, 2.029, and 4.913. We check the endpoints and critical points for the global maximum and minimum.

$$f(0) = 0, \qquad\qquad f(2\pi) = 0,$$
$$f(2.029) \approx 1.8197, \qquad f(4.914) \approx -4.814.$$

Thus for $0 \le x \le 2\pi$, $-4.81 \le f(x) \le 1.82$.

37. (a) We have $g'(t) = \frac{t(1/t) - \ln t}{t^2} = \frac{1 - \ln t}{t^2}$, which is zero if $t = e$, negative if $t > e$, and positive if $t < e$, since $\ln t$ is increasing. Thus $g(e) = \frac{1}{e}$ is a global maximum for g. Since $t = e$ was the only point at which $g'(t) = 0$, there is no minimum.

(b) Now $\ln t/t$ is increasing for $0 < t < e$, $\ln 1/1 = 0$, and $\ln 5/5 \approx 0.322 < \ln(e)/e$. Thus, for $1 < t < e$, $\ln t/t$ increases from 0 to above $\ln 5/5$, so there must be a t between 1 and e such that $\ln t/t = \ln 5/5$. For $t > e$, there is only one solution to $\ln t/t = \ln 5/5$, namely $t = 5$, since $\ln t/t$ is decreasing for $t > e$. For $0 < t < 1$, $\ln t/t$ is negative and so cannot equal $\ln 5/5$. Thus $\ln x/x = \ln t/t$ has exactly two solutions.

(c) The graph of $\ln t/t$ intersects the horizontal line $y = \ln 5/5$, at $x = 5$ and $x \approx 1.75$.

CAS Challenge Problems

41. (a) Since $k > 0$, we have $\lim\limits_{t \to \infty} e^{-kt} = 0$. Thus

$$\lim_{t \to \infty} P = \lim_{t \to \infty} \frac{L}{1 + Ce^{-kt}} = \frac{L}{1 + C \cdot 0} = L.$$

The constant L is called the carrying capacity of the environment because it represents the long-run population in the environment.

(b) Using a CAS, we find

$$\frac{d^2 P}{dt^2} = -\frac{LCk^2 e^{-kt}(1 - Ce^{-kt})}{(1 + Ce^{-kt})^3}.$$

Thus, $d^2P/dt^2 = 0$ when

$$1 - Ce^{-kt} = 0$$
$$t = -\frac{\ln(1/C)}{k}.$$

Since e^{-kt} and $(1 + Ce^{-kt})$ are both always positive, the sign of d^2P/dt^2 is negative when $(1 - Ce^{-kt}) > 0$, that is, for $t > -\ln(1/C)/k$. Similarly, the sign of d^2P/dt^2 is positive when $(1 - Ce^{-kt}) < 0$, that is, for $t < -\ln(1/C)/k$. Thus, there is an inflection point at $t = -\ln(1/C)/k$.
For $t = -\ln(1/C)/k$,

$$P = \frac{L}{1 + Ce^{\ln(1/C)}} = \frac{L}{1 + C(1/C)} = \frac{L}{2}.$$

Thus, the inflection point occurs where $P = L/2$.

45. (a) Using a computer algebra system or differentiating by hand, we get

$$f'(x) = \frac{1}{2\sqrt{a+x}(\sqrt{a}+\sqrt{x})} - \frac{\sqrt{a+x}}{2\sqrt{x}(\sqrt{a}+\sqrt{x})^2}.$$

Simplifying gives

$$f'(x) = \frac{-a + \sqrt{a}\sqrt{x}}{2\left(\sqrt{a}+\sqrt{x}\right)^2 \sqrt{x}\sqrt{a+x}}.$$

The denominator of the derivative is always positive if $x > 0$, and the numerator is zero when $x = a$. Writing the numerator as $\sqrt{a}(\sqrt{x} - \sqrt{a})$, we see that the derivative changes from negative to positive at $x = a$. Thus, by the first derivative test, the function has a local minimum at $x = a$.

(b) As a increases, the local minimum moves to the right. See Figure 4.21. This is consistent with what we found in part (a), since the local minimum is at $x = a$.

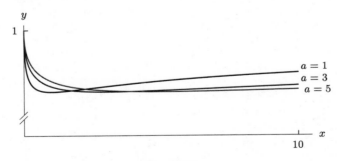

Figure 4.21

(c) Using a computer algebra system to find the second derivative when $a = 2$, we get

$$f''(x) = \frac{4\sqrt{2} + 12\sqrt{x} + 6x^{3/2} - 3\sqrt{2}x^2}{4\left(\sqrt{2}+\sqrt{x}\right)^3 x^{3/2}(2+x)^{3/2}}.$$

Using the computer algebra system again to solve $f''(x) = 0$, we find that it has one zero at $x = 4.6477$. Graphing the second derivative, we see that it goes from positive to negative at $x = 4.6477$, so this is an inflection point.

CHECK YOUR UNDERSTANDING

1. True. Since the domain of f is all real numbers, all local minima occur at critical points.

5. False. For example, if $f(x) = x^3$, then $f'(0) = 0$, but $f(x)$ does not have either a local maximum or a local minimum at $x = 0$.

9. True, by the Increasing Function Theorem, Theorem 4.4.

13. Let $f(x) = ax^2$, with $a \neq 0$. Then $f'(x) = 2ax$, so f has a critical point only at $x = 0$.

17. Let f be defined by

$$f(x) = \begin{cases} x^2 & \text{if } 0 \le x < 1 \\ 1/2 & \text{if } x = 1. \end{cases}$$

Then f is not continuous at $x = 1$, but f is differentiable on $(0, 1)$ and $f'(x) = 2x$ for $0 < x < 1$. Thus, $c = 1/4$ satisfies

$$f'(c) = \frac{f(1) - f(0)}{1 - 0} = \frac{1}{2}, \quad \text{since} \quad f'\left(\frac{1}{4}\right) = 2 \cdot \frac{1}{4} = \frac{1}{2}.$$

21. This is impossible. If $f(a) > 0$, then the downward concavity forces the graph of f to cross the x-axis to the right or left of $x = a$, which means $f(x)$ cannot be positive for all values of x. More precisely, suppose that $f(x)$ is positive for all x and f is concave down. Thus there must be some value $x = a$ where $f(a) > 0$ and $f'(a)$ is not zero, since a constant function is not concave down. The tangent line at $x = a$ has nonzero slope and hence must cross the x-axis somewhere to the right or left of $x = a$. Since the graph of f must lie below this tangent line, it must also cross the x-axis, contradicting the assumption that $f(x)$ is positive for all x.

25. This is impossible. Since f''' exists, f'' must be continuous. By the Intermediate Value Theorem, $f''(x)$ cannot change sign, since $f''(x)$ cannot be zero. In the same way, we can show that $f'(x)$ and $f(x)$ cannot change sign. Since the product of these three with $f'''(x)$ cannot change sign, $f'''(x)$ cannot change sign. Thus $f(x)f''(x)$ and $f'(x)f'''(x)$ cannot change sign. Since their product is negative for all x, one or the other must be negative for all x. By Problem 24, this is impossible.

CHAPTER FIVE

Solutions for Section 5.1

Exercises

1. (a) Suppose $f(t)$ is the flowrate in m^3/hr at time t. We are only given two values of the flowrate, so in making our estimates of the flow, we use one subinterval, with $\Delta t = 3/1 = 3$:

$$\text{Left estimate} = 3[f(6 \text{ am})] = 3 \cdot 100 = 300 \text{ m}^3 \quad \text{(an underestimate)}$$

$$\text{Right estimate} = 3[f(9 \text{ am})] = 3 \cdot 280 = 840 \text{ m}^3 \quad \text{(an overestimate)}.$$

The best estimate is the average of these two estimates,

$$\text{Best estimate} = \frac{\text{Left} + \text{Right}}{2} = \frac{300 + 840}{2} = 570 \text{ m}^3.$$

(b) Since the flowrate is increasing throughout, the error, i.e., the difference between over- and under-estimates, is given by

$$\text{Error} \leq \Delta t\,[f(9 \text{ am}) - f(6 \text{ am})] = \Delta t[280 - 100] = 180\Delta t.$$

We wish to choose Δt so that the the error $180\Delta t \leq 6$, or $\Delta t \leq 6/180 = 1/30$. So the flowrate guage should be read every $1/30$ of an hour, or every 2 minutes.

Problems

5. (a) Car A has the largest maximum velocity because the peak of car A's velocity curve is higher than the peak of B's.
   (b) Car A stops first because the curve representing its velocity hits zero (on the t-axis) first.
   (c) Car B travels farther because the area under car B's velocity curve is the larger.

9. Using $\Delta t = 0.2$, our upper estimate is

$$\frac{1}{1+0}(0.2) + \frac{1}{1+0.2}(0.2) + \frac{1}{1+0.4}(0.2) + \frac{1}{1+0.6}(0.2) + \frac{1}{1+0.8}(0.2) \approx 0.75.$$

The lower estimate is

$$\frac{1}{1+0.2}(0.2) + \frac{1}{1+0.4}(0.2) + \frac{1}{1+0.6}(0.2) + \frac{1}{1+0.8}(0.2)\frac{1}{1+1}(0.2) \approx 0.65.$$

Since v is a decreasing function, the bug has crawled more than 0.65 meters, but less than 0.75 meters. We average the two to get a better estimate:

$$\frac{0.65 + 0.75}{2} = 0.70 \text{ meters.}$$

13. (a) An upper estimate is $9.81 + 8.03 + 6.53 + 5.38 + 4.41 = 34.16$ m/sec. A lower estimate is $8.03 + 6.53 + 5.38 + 4.41 + 3.61 = 27.96$ m/sec.
   (b) The average is $\frac{1}{2}(34.16 + 27.96) = 31.06$ m/sec. Because the graph of acceleration is concave up, this estimate is too high, as can be seen in the figure to below. The area of the shaded region is the average of the areas of the rectangles $ABFE$ and $CDFE$.

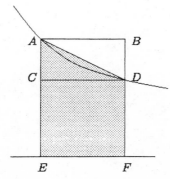

Solutions for Section 5.2

Exercises

1.

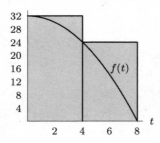

Figure 5.1: Left Sum, $\Delta t = 4$

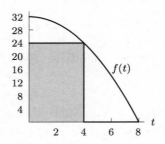

Figure 5.2: Right Sum, $\Delta t = 4$

(a) Left-hand sum $= 32(4) + 24(4) = 224$.
(b) Right-hand sum $= 24(4) + 0(4) = 96$.

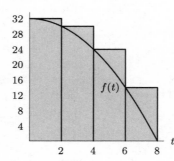

Figure 5.3: Left Sum, $\Delta t = 2$

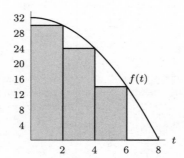

Figure 5.4: Right Sum, $\Delta t = 2$

(c) Left-hand sum $= 32(2) + 30(2) + 24(2) + 14(2) = 200$.
(d) Right-hand sum $= 30(2) + 24(2) + 14(2) + 0(2) = 136$.

5. With $\Delta x = 5$, we have

$$\text{Left-hand sum} = 5(0 + 100 + 200 + 100 + 200 + 250 + 275) = 5625,$$

$$\text{Right-hand sum} = 5(100 + 200 + 100 + 200 + 250 + 275 + 300) = 7125.$$

The average of these two sums is our best guess for the value of the integral;

$$\int_{-15}^{20} f(x)\, dx \approx \frac{5625 + 7125}{2} = 6375.$$

9. We use a calculator or computer to see that $\int_{0}^{3} 2^x dx = 10.0989$.

13.

n	2	10	50	250
Left-hand Sum	1.34076	1.07648	1.01563	1.00314
Right-hand Sum	0.55536	0.91940	0.98421	0.99686

The sums seem to be converging to 1. Since $\cos x$ is monotone on $[0, \pi/2]$, the true value is between 1.00314 and .99686 .

17. The graph of $y = 7 - x^2$ has intercepts $x = \pm\sqrt{7}$. See Figure 5.5. Therefore we have

$$\text{Area} = \int_{-\sqrt{7}}^{\sqrt{7}} (7 - x^2)\, dx = 24.7.$$

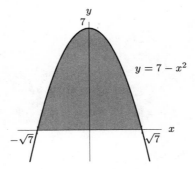

Figure 5.5

21. Since $x^{1/2} \le x^{1/3}$ for $0 \le x \le 1$, we have

$$\text{Area} = \int_{0}^{1} (x^{1/3} - x^{1/2})\, dx = 0.0833.$$

Problems

25. The areas we computed are shaded in Figure 5.6. Since $y = x^2$ and $y = x^{1/2}$ are inverse functions, their graphs are reflections about the line $y = x$. Similarly, $y = x^3$ and $y = x^{1/3}$ are inverse functions and their graphs are reflections about the line $y = x$. Therefore, the two shaded areas in Figure 5.6 are equal.

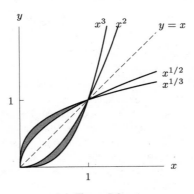

Figure 5.6

29. We have $\Delta x = 2/500 = 1/250$. The formulas for the left- and right-hand Riemann sums give us that

$$\text{Left} = \Delta x[f(-1) + f(-1 + \Delta x) + \ldots + f(1 - 2\Delta x) + f(1 - \Delta x)]$$
$$\text{Right} = \Delta x[f(-1 + \Delta x) + f(-1 + 2\Delta x) + \ldots + f(1 - \Delta x) + f(1)].$$

Subtracting these yields

$$\text{Right} - \text{Left} = \Delta x[f(1) - f(-1)] = \frac{1}{250}[6 - 2] = \frac{4}{250} = \frac{2}{125}.$$

33. We have

$$\Delta x = \frac{4}{3} = \frac{b-a}{n} \quad \text{and} \quad n = 3, \quad \text{so} \quad b - a = 4 \quad \text{or} \quad b = a + 4.$$

The function, $f(x)$, is squaring something. Since it is a left-hand sum, $f(x)$ could equal x^2 with $a = 2$ and $b = 6$ (note that $2 + 3(\frac{4}{3})$ gives the right-hand endpoint of the last interval). Or, $f(x)$ could possibly equal $(x + 2)^2$ with $a = 0$ and $b = 4$. Other answers are possible.

37.

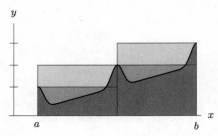

Figure 5.7: Integral vs. Left- and Right-Hand Sums

Solutions for Section 5.3

Exercises

1. Average value $= \dfrac{1}{2 - 0} \displaystyle\int_0^2 (1 + t)\, dt = \dfrac{1}{2}(4) = 2.$

5. The units of measurement are dollars.

9. (a)

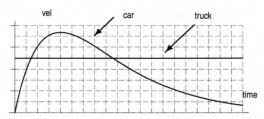

(b) The graphs intersect twice, at about 0.7 hours and 4.3 hours. At each intersection point, the velocity of the car is equal to the velocity of the truck, so $v_{\text{car}} = v_{\text{truck}}$. From the time they start until 0.7 hours later, the truck is traveling at a greater velocity than the car, so the truck is ahead of the car and is pulling farther away. At 0.7 hours they are traveling at the same velocity, and after 0.7 hours the car is traveling faster than the truck, so that the car begins to gain on the truck. Thus, at 0.7 hours the truck is farther from the car than it is immediately before or after 0.7 hours.

Similarly, because the car's velocity is greater than the truck's after 0.7 hours, it will catch up with the truck and eventually pass and pull away from the truck until 4.3 hours, at which point the two are again traveling at the same velocity. After 4.3 hours the truck travels faster than the car, so that it now gains on the car. Thus, 4.3 hours represents the point where the car is farthest ahead of the truck.

Problems

13. (a) The integral is the area above the x-axis minus the area below the x-axis. Thus, we can see that $\int_{-3}^{3} f(x)\, dx$ is about $-6 + 2 = -4$ (the negative of the area from $t = -3$ to $t = 1$ plus the area from $t = 1$ to $t = 3$.)

(b) Since the integral in part (a) is negative, the average value of $f(x)$ between $x = -3$ and $x = 3$ is negative. From the graph, however, it appears that the average value of $f(x)$ from $x = 0$ to $x = 3$ is positive. Hence (ii) is the larger quantity.

17. The time period 9am to 5pm is represented by the time $t = 0$ to $t = 8$ and $t = 24$ to $t = 32$. The area under the curve, or total number of worker-hours for these times, is about 9 boxes or $9(80) = 720$ worker-hours. The total cost for 9am to 5pm is $(720)(10) = \$7200$. The area under the rest of the curve is about 5.5 boxes, or $5.5(80) = 440$ worker-hours. The total cost for this time period is $(440)(15) = \$6600$. The total cost is about $7200 + 6600 = \$13,800$.

21. Since the average value is given by

$$\text{Average value} = \frac{1}{b-a} \int_a^b f(x)\,dx,$$

the units for dx inside the integral are canceled by the units for $1/(b-a)$ outside the integral, leaving only the units for $f(x)$. This is as it should be, since the average value of f should be measured in the same units as $f(x)$.

25. Change in income $= \int_0^{12} r(t)\,dt = \int_0^{12} 40(1.002)^t\,dt = \$485.80$

29. We know that the the integral of F, and therefore the work, can be obtained by computing the areas in Figure 5.8.

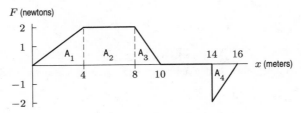

Figure 5.8

$$W = \int_0^{16} F(x)\,dx = \text{Area above } x\text{-axis} - \text{Area below } x\text{-axis}$$

$$= A_1 + A_2 + A_3 - A_4$$

$$= \frac{1}{2} \cdot 4 \cdot 2 + 4 \cdot 2 + \frac{1}{2} \cdot 2 \cdot 2 - \frac{1}{2} \cdot 2 \cdot 2$$

$$= 12 \text{ newton} \cdot \text{meters.}$$

Solutions for Section 5.4

Exercises

1. We find the changes in $f(x)$ between any two values of x by counting the area between the curve of $f'(x)$ and the x-axis. Since $f'(x)$ is linear throughout, this is quite easy to do. From $x = 0$ to $x = 1$, we see that $f'(x)$ outlines a triangle of area $1/2$ below the x-axis (the base is 1 and the height is 1). By the Fundamental Theorem,

$$\int_0^1 f'(x)\,dx = f(1) - f(0),$$

so

$$f(0) + \int_0^1 f'(x)\,dx = f(1)$$

$$f(1) = 2 - \frac{1}{2} = \frac{3}{2}$$

Similarly, between $x = 1$ and $x = 3$ we can see that $f'(x)$ outlines a rectangle below the x-axis with area -1, so $f(2) = 3/2 - 1 = 1/2$. Continuing with this procedure (note that at $x = 4$, $f'(x)$ becomes positive), we get the table below.

x	0	1	2	3	4	5	6
$f(x)$	2	3/2	1/2	$-1/2$	-1	$-1/2$	1/2

Problems

5. Note that $\int_a^b f(z)\,dz = \int_a^b f(x)\,dx$. Thus, we have

$$\int_a^b cf(z)\,dz = c\int_a^b f(z)\,dz = 8c.$$

9. The graph of $y = f(x-5)$ is the graph of $y = f(x)$ shifted to the right by 5. Since the limits of integration have also shifted by 5 (to $a+5$ and $b+5$), the areas corresponding to $\int_{a+5}^{b+5} f(x-5)\,dx$ and $\int_a^b f(x)\,dx$ are the same. Thus,

$$\int_{a+5}^{b+5} f(x-5)\,dx = \int_a^b f(x)\,dx = 8.$$

13. (a) The integrand is positive, so the integral can't be negative.
   (b) The integrand ≥ 0. If the integral $= 0$, then the integrand must be identically 0, which isn't true.

17.

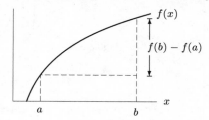

21. (a) $\dfrac{1}{\sqrt{2\pi}}\displaystyle\int_1^3 e^{-\frac{x^2}{2}}\,dx$

$$= \frac{1}{\sqrt{2\pi}}\int_0^3 e^{-\frac{x^2}{2}}\,dx - \frac{1}{\sqrt{2\pi}}\int_0^1 e^{-\frac{x^2}{2}}\,dx$$
$$\approx 0.4987 - 0.3413 = 0.1574.$$

   (b) $\left(\text{by symmetry of } e^{x^2/2}\right)$ $\dfrac{1}{\sqrt{2\pi}}\displaystyle\int_{-2}^3 e^{-\frac{x^2}{2}}\,dx = \frac{1}{\sqrt{2\pi}}\int_{-2}^0 e^{-\frac{x^2}{2}}\,dx + \frac{1}{\sqrt{2\pi}}\int_0^3 e^{-\frac{x^2}{2}}\,dx$

$$= \frac{1}{\sqrt{2\pi}}\int_0^2 e^{-\frac{x^2}{2}}\,dx + \frac{1}{\sqrt{2\pi}}\int_0^3 e^{-\frac{x^2}{2}}\,dx$$
$$\approx 0.4772 + 0.4987 = 0.9759.$$

Solutions for Chapter 5 Review

Exercises

1. (a) We calculate the right- and left-hand sums as follows:

$$\text{Left} = 2[80 + 52 + 28 + 10] = 340 \text{ ft.}$$
$$\text{Right} = 2[52 + 28 + 10 + 0] = 180 \text{ ft.}$$

Our best estimate will be the average of these two sums,

$$\text{Best} = \frac{\text{Left} + \text{Right}}{2} = \frac{340 + 180}{2} = 260 \text{ ft.}$$

   (b) Since v is decreasing throughout,

$$\text{Left} - \text{Right} = \Delta t \cdot [f(0) - f(8)]$$
$$= 80\Delta t.$$

Since our best estimate is the average of Left and Right, the maximum error is $(80)\Delta t/2$. For $(80)\Delta t/2 \leq 20$, we must have $\Delta t \leq 1/2$. In other words, we must measure the velocity every 0.5 second.

5. We take $\Delta t = 20$. Then:

$$\text{Left-hand sum} = 1.2(20) + 2.8(20) + 4.0(20) + 4.7(20) + 5.1(20)$$
$$= 356.$$
$$\text{Right-hand sum} = 2.8(20) + 4.0(20) + 4.7(20) + 5.1(20) + 5.2(20)$$
$$= 436.$$
$$\int_0^{100} f(t)\,dt \approx \text{Average} = \frac{356 + 436}{2} = 396.$$

9. The x intercepts of $y = x^2 - 9$ are $x = -3$ and $x = 3$, and since the graph is below the x axis on the interval $[-3, 3]$.

$$\text{Area} = -\int_{-3}^{3} (x^2 - 9)\,dx = 36.00.$$

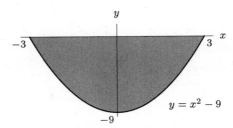

13. The graph of $y = -x^2 + 5x - 4$ is shown in Figure 5.9. We wish to find the area shaded. Since the graph crosses the x-axis at $x = 1$, we must split the integral at $x = 1$. For $x < 1$, the graph is below the x-axis, so the area is the negative of the integral. Thus

$$\text{Area shaded} = -\int_0^1 (-x^2 + 5x - 4)\,dx + \int_1^3 (-x^2 + 5x - 4)\,dx.$$

Using a calculator or computer, we find

$$\int_0^1 (-x^2 + 5x - 4)\,dx = -1.8333 \quad \text{and} \quad \int_1^3 (-x^2 + 5x - 4)\,dx = 3.3333.$$

Thus,

$$\text{Area shaded} = 1.8333 + 3.3333 = 5.1666.$$

(Notice that $\int_0^3 f(x)\,dx = -1.8333 + 3.333 = 1.5$, but the value of this integral is not the area shaded.)

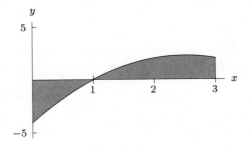

Figure 5.9

Problems

17. From $t = 0$ to $t = 3$, you are moving away from home ($v > 0$); thereafter you move back toward home. So you are the farthest from home at $t = 3$. To find how far you are then, we can measure the area under the v curve as about 9 squares, or $9 \cdot 10$ km/hr $\cdot$ 1 hr $= 90$ km. To find how far away from home you are at $t = 5$, we measure the area from $t = 3$ to $t = 5$ as about 25 km, except that this distance is directed toward home, giving a total distance from home during the trip of $90 - 25 = 65$ km.

21. (a) Clearly, the points where $x = \sqrt{\pi}, \sqrt{2\pi}, \sqrt{3\pi}, \sqrt{4\pi}$ are where the graph intersects the x-axis because $f(x) = \sin(x^2) = 0$ where x is the square root of some multiple of π.

(b) Let $f(x) = \sin(x^2)$, and let A, B, C, and D be the areas of the regions indicated in the figure below. Then we see that $A > B > C > D$.

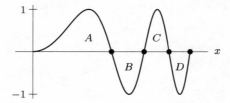

Note that

$$\int_0^{\sqrt{\pi}} f(x)\,dx = A, \qquad \int_0^{\sqrt{2\pi}} f(x)\,dx = A - B,$$

$$\int_0^{\sqrt{3\pi}} f(x)\,dx = A - B + C, \quad \text{and} \quad \int_0^{\sqrt{4\pi}} f(x)\,dx = A - B + C - D.$$

It follows that

$$\int_0^{\sqrt{\pi}} f(x)\,dx = A > \int_0^{\sqrt{3\pi}} f(x)\,dx = A - (B - C) = A - B + C >$$

$$\int_0^{\sqrt{4\pi}} f(x)\,dx = A - B + C - D > \int_0^{\sqrt{2\pi}} f(x)\,dx = (A - B) > 0.$$

And thus the ordering is $n = 1$, $n = 3$, $n = 4$, and $n = 2$ from largest to smallest. All the numbers are positive.

25. (a) We know that $\int_2^5 f(x)\,dx = \int_0^5 f(x)\,dx - \int_0^2 f(x)\,dx$. By symmetry, $\int_0^2 f(x)\,dx = \frac{1}{2} \int_{-2}^2 f(x)\,dx$, so $\int_2^5 f(x)\,dx = \int_0^5 f(x)\,dx - \frac{1}{2} \int_{-2}^2 f(x)\,dx$.

(b) $\int_2^5 f(x)\,dx = \int_{-2}^5 f(x)\,dx - \int_{-2}^2 f(x)\,dx = \int_{-2}^5 f(x)\,dx - 2\int_{-2}^0 f(x)\,dx$.

(c) Using symmetry again, $\int_0^2 f(x)\,dx = \frac{1}{2} \left(\int_{-2}^5 f(x)\,dx - \int_2^5 f(x)\,dx \right)$.

29. The change in the amount of water is the integral of rate of change, so we have

$$\text{Number of liters pumped out} = \int_0^{60} (5 - 5e^{-0.12t})\,dt = 258.4 \text{ liters}.$$

Since the tank contained 1000 liters of water initially, we see that

$$\text{Amount in tank after one hour} = 1000 - 258.4 = 741.6 \text{ liters}.$$

33. (a) About 300 meter^3/sec.

(b) About 250 meter^3/sec.

(c) Looking at the graph, we can see that the 1996 flood reached its maximum just between March and April, for a high of about 1250 meter^3/sec. Similarly, the 1957 flood reached its maximum in mid-June, for a maximum flow rate of 3500 meter^3/sec.

(d) The 1996 flood lasted about 1/3 of a month, or about 10 days. The 1957 flood lasted about 4 months.

(e) The area under the controlled flood graph is about 2/3 box. Each box represents 500 meter^3/sec for one month. Since

$$1 \text{ month} = 30\frac{\text{days}}{\text{month}} \cdot 24\frac{\text{hours}}{\text{day}} \cdot 60\frac{\text{minutes}}{\text{hour}} \cdot 60\frac{\text{seconds}}{\text{minute}}$$

$$= 2.592 \cdot 10^6 \approx 3 \cdot 10^6 \text{seconds},$$

each box represents

$$\text{Flow} \approx (500 \text{ meter}^3/\text{sec}) \cdot (2.6 \cdot 10^6 \text{ sec}) = 13 \cdot 10^8 \text{ meter}^3\text{of water}.$$

So, for the artificial flood,

$$\text{Additional flow} \approx \frac{2}{3} \cdot 13 \cdot 10^8 = 9 \cdot 10^8 \text{ meter}^3 \approx 10^9 \text{ meter}^3.$$

(f) The 1957 flood released a volume of water represented by about 12 boxes above the 250 meter/sec baseline. Thus, for the natural flood,

$$\text{Additional flow} \approx 12 \cdot 15 \cdot 10^8 = 1.8 \cdot 10^{10} \approx 2 \cdot 10^{10} \text{ meter}^3.$$

So, the natural flood was nearly 20 times larger than the controlled flood and lasted much longer.

37. The graph of rate against time is the straight line shown in Figure 5.10. Since the shaded area is 270, we have

$$\frac{1}{2}(10 + 50) \cdot t = 270$$

$$t = \frac{270}{60} \cdot 2 = 9 \text{ years}$$

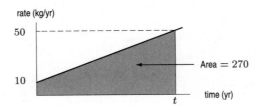

Figure 5.10

CAS Challenge Problems

41. (a) Since the length of the interval of integration is $2 - 1 = 1$, the width of each subdivision is $\Delta t = 1/n$. Thus the endpoints of the subdivision are

$$t_0 = 1, \quad t_1 = 1 + \Delta t = 1 + \frac{1}{n}, \quad t_2 = 1 + 2\Delta t = 1 + \frac{2}{n}, \ldots,$$

$$t_i = 1 + i\Delta t = 1 + \frac{i}{n}, \ldots, \quad t_{n-1} = 1 + (n-1)\Delta t = 1 + \frac{n-1}{n}.$$

Thus, since the integrand is $f(t) = t^2$,

$$\text{Left-hand sum} = \sum_{i=0}^{n-1} f(t_i)\Delta t = \sum_{i=0}^{n-1} t_i^2 \Delta t = \sum_{i=0}^{n-1} \left(1 + \frac{i}{n}\right)^2 \frac{1}{n} = \sum_{i=0}^{n-1} \frac{(n+i)^2}{n^3}.$$

(b) Using a CAS to find the sum, we get

$$\sum_{i=0}^{n-1} \frac{(n+i)^2}{n^3} = \frac{(-1 + 2n)(-1 + 7n)}{6n^2} = \frac{7}{3} + \frac{1}{6n^2} - \frac{3}{2n}.$$

(c) Taking the limit as $n \to \infty$

$$\lim_{n\to\infty} \left(\frac{7}{3} + \frac{1}{6n^2} - \frac{3}{2n}\right) = \lim_{n\to\infty} \frac{7}{3} + \lim_{n\to\infty} \frac{1}{6n^2} - \lim_{n\to\infty} \frac{3}{2n} = \frac{7}{3} + 0 + 0 = \frac{7}{3}.$$

(d) We have calculated $\int_1^2 t^2\, dt$ using Riemann sums. Since t^2 is above the t-axis between $t = 1$ and $t = 2$, this integral is the area; so the area is 7/3.

CHECK YOUR UNDERSTANDING

1. True, since $\int_0^2 (f(x) + g(x))dx = \int_0^2 f(x)dx + \int_0^2 g(x)dx$.

5. False. This would be true if $h(x) = 5f(x)$. However, we cannot assume that $f(5x) = 5f(x)$, so for many functions this statement is false. For example, if f is the constant function $f(x) = 3$, then $h(x) = 3$ as well, so $\int_0^2 f(x)\, dx = \int_0^2 h(x)\, dx = 6$.

9. **False.** If the graph of f is symmetric about the y-axis, this is true, but otherwise it is usually not true. For example, if $f(x) = x + 1$ the area under the graph of f for $-1 \leq x \leq 0$ is less than the area under the graph of f for $0 \leq x \leq 1$, so $\int_{-1}^{1} f(x)\,dx < 2 \int_{0}^{1} f(x)\,dx$. See Figure 5.11.

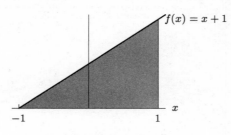

Figure 5.11

13. **True**, by Theorem 5.4 on Comparison of Definite Integrals:

$$\frac{1}{b-a} \int_{a}^{b} f(x)\,dx \leq \frac{1}{b-a} \int_{a}^{b} g(x)\,dx.$$

17. **True.** By the properties of integrals in Theorem 5.3, we have:

$$\int_{a}^{b} (f(x) + g(x))dx = \int_{a}^{b} f(x)dx + \int_{a}^{b} g(x)dx.$$

Dividing both sides of this equation through by $b - a$, we get that the average value of $f(x) + g(x)$ is average value of $f(x)$ plus the average value of $g(x)$:

$$\frac{1}{b-a} \int_{a}^{b} (f(x) + g(x))\,dx = \frac{1}{b-a} \int_{a}^{b} f(x)\,dx + \frac{1}{b-a} \int_{a}^{b} g(x)\,dx.$$

CHAPTER SIX

Solutions for Section 6.1

Exercises

1.

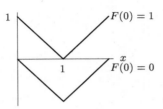

5. By the Fundamental Theorem of Calculus, we know that

$$f(2) - f(0) = \int_0^2 f'(x)dx.$$

Using a left-hand sum, we estimate $\int_0^2 f'(x)dx \approx (10)(2) = 20$. Using a right-hand sum, we estimate $\int_0^2 f'(x)dx \approx (18)(2) = 36$. Averaging, we have

$$\int_0^2 f'(x)dx \approx \frac{20 + 36}{2} = 28.$$

We know $f(0) = 100$, so

$$f(2) = f(0) + \int_0^2 f'(x)dx \approx 100 + 28 = 128.$$

Similarly, we estimate

$$\int_2^4 f'(x)dx \approx \frac{(18)(2) + (23)(2)}{2} = 41,$$

so

$$f(4) = f(2) + \int_2^4 f'(x)dx \approx 128 + 41 = 169.$$

Similarly,

$$\int_4^6 f'(x)dx \approx \frac{(23)(2) + (25)(2)}{2} = 48,$$

so

$$f(6) = f(4) + \int_4^6 f'(x)dx \approx 169 + 48 = 217.$$

The values are shown in the table.

x	0	2	4	6
$f(x)$	100	128	169	217

Problems

9. (a) Critical points of $F(x)$ are the zeros of f: $x = 1$ and $x = 3$.
   (b) $F(x)$ has a local minimum at $x = 1$ and a local maximum at $x = 3$.
   (c)

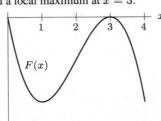

Notice that the graph could also be above or below the x-axis at $x = 3$.

13.

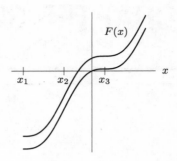

Note that since $f(x_1) = 0$, $F(x_1)$ is either a local minimum or a point of inflection; it is impossible to tell which from the graph. Since $f'(x_3) = 0$, and f' changes sign around $x = x_3$, $F(x_3)$ is an inflection point. Also, since $f'(x_2) = 0$ and f changes from increasing to decreasing about $x = x_2$, F has another inflection point at $x = x_2$.

17. The critical points are at $(0, 5)$, $(2, 21)$, $(4, 13)$, and $(5, 15)$. A graph is given below.

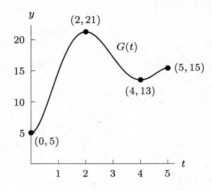

21. (a) The total volume emptied must increase with time and cannot decrease. The smooth graph (I) that is always increasing is therefore the volume emptied from the bladder. The jagged graph (II) that increases then decreases to zero is the flow rate.

(b) The total change in volume is the integral of the flow rate. Thus, the graph giving total change (I) shows an antiderivative of the rate of change in graph (II).

25. (a) Suppose $Q(t)$ is the amount of water in the reservoir at time t. Then

$$Q'(t) = \begin{array}{c} \text{Rate at which water} \\ \text{in reservoir is changing} \end{array} = \begin{array}{c} \text{Inflow} \\ \text{rate} \end{array} - \begin{array}{c} \text{Outflow} \\ \text{rate} \end{array}$$

Thus the amount of water in the reservoir is increasing when the inflow curve is above the outflow, and decreasing when it is below. This means that $Q(t)$ is a maximum where the curves cross in July 1993 (as shown in Figure 6.1), and $Q(t)$ is decreasing fastest when the outflow is farthest above the inflow curve, which occurs about October 1993 (see Figure 6.1).

To estimate values of $Q(t)$, we use the Fundamental Theorem which says that the change in the total quantity of water in the reservoir is given by

$$Q(t) - Q(\text{Jan'93}) = \int_{\text{Jan93}}^{t} (\text{inflow rate} - \text{outflow rate}) \, dt$$

or $\qquad Q(t) = Q(\text{Jan'93}) + \int_{\text{Jan93}}^{t} (\text{inflow rate} - \text{outflow rate}) \, dt.$

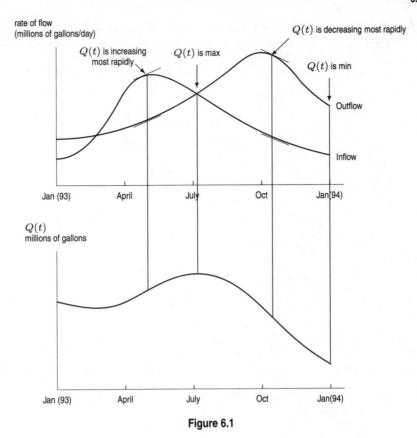

Figure 6.1

(b) See Figure 6.1. Maximum in July 1993. Minimum in Jan 1994.
(c) See Figure 6.1. Increasing fastest in May 1993. Decreasing fastest in Oct 1993.
(d) In order for the water to be the same as Jan '93 the total amount of water which has flowed into the reservoir must be 0. Referring to Figure 6.2, we have

$$\int_{Jan93}^{July94} (\text{inflow} - \text{outflow})dt = -A_1 + A_2 - A_3 + A_4 = 0$$

giving $A_1 + A_3 = A_2 + A_4$

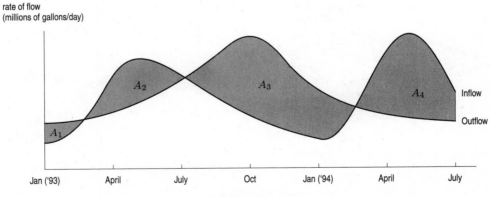

Figure 6.2

Solutions for Section 6.2

Exercises

1. $5x$

5. $\sin t$

9. $-\dfrac{1}{2z^2}$

13. $\dfrac{t^4}{4} - \dfrac{t^3}{6} - \dfrac{t^2}{2}$

17. $-\cos 2\theta$

21. $\dfrac{5}{2}x^2 - \dfrac{2}{3}x^{\frac{3}{2}}$

25. $R(t) = \displaystyle\int (t^3 + 5t - 1)\, dt = \dfrac{t^4}{4} + \dfrac{5}{2}t^2 - t + C$

29. $H(x) = \displaystyle\int (4x^3 - 7)\, dx = x^4 - 7x + C$

33. $F(x) = \displaystyle\int \dfrac{1}{x^2}\, dx = -\dfrac{1}{x} + C$

37. $f(x) = \frac{1}{4}x$, so $F(x) = \frac{x^2}{8} + C$. $F(0) = 0$ implies that $\frac{1}{8} \cdot 0^2 + C = 0$, so $C = 0$. Thus $F(x) = x^2/8$ is the only possibility.

41. $f(x) = \sin x$, so $F(x) = -\cos x + C$. $F(0) = 0$ implies that $-\cos 0 + C = 0$, so $C = 1$. Thus $F(x) = -\cos x + 1$ is the only possibility.

45. $\displaystyle\int (x^3 - 2)\, dx = \dfrac{x^4}{4} - 2x + C$

49. $\displaystyle\int \dfrac{4}{t^2}\, dt = -\dfrac{4}{t} + C$

53. $\dfrac{x^2}{2} + 2x^{1/2} + C$

57. $\sin(x + 1) + C$

61. $\displaystyle\int_0^3 (x^2 + 4x + 3)\, dx = \left(\dfrac{x^3}{3} + 2x^2 + 3x\right)\Big|_0^3 = (9 + 18 + 9) - 0 = 36$

65. $\displaystyle\int_2^5 (x^3 - \pi x^2)\, dx = \left(\dfrac{x^4}{4} - \dfrac{\pi x^3}{3}\right)\Big|_2^5 = \dfrac{609}{4} - 39\pi \approx 29.728.$

69. $\displaystyle\int_0^{\pi/4} (\sin t + \cos t)\, dt = (-\cos t + \sin t)\Big|_0^{\pi/4} = \left(-\dfrac{\sqrt{2}}{2} + \dfrac{\sqrt{2}}{2}\right) - (-1 + 0) = 1.$

73. $\displaystyle\int 2^x\, dx = \dfrac{1}{\ln 2} 2^x + C$, since $\dfrac{d}{dx} 2^x = \ln 2 \cdot 2^x$, so

$\displaystyle\int_{-1}^1 2^x\, dx = \dfrac{1}{\ln 2}\left[2^x \Big|_{-1}^1\right] = \dfrac{3}{2\ln 2} \approx 2.164.$

Problems

77. Since the graph of $y = e^x$ is above the graph of $y = \cos x$ (see the figure below), we have

$$\text{Area} = \int_0^1 (e^x - \cos x)\, dx$$

$$= \int_0^1 e^x\, dx - \int_0^1 \cos x\, dx$$

$$= e^x \Big|_0^1 - \sin x \Big|_0^1$$

$$= e^1 - e^0 - \sin 1 + \sin 0$$

$$= e - 1 - \sin 1.$$

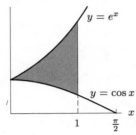

81. The average value of $v(x)$ on the interval $1 \le x \le c$ is

$$\frac{1}{c-1} \int_1^c \frac{6}{x^2}\, dx = \frac{1}{c-1}\left(-\frac{6}{x}\right)\Big|_1^c = \frac{1}{c-1}\left(\frac{-6}{c} + 6\right) = \frac{6}{c}.$$

Since $\dfrac{1}{c-1} \displaystyle\int_1^c \dfrac{6}{x^2}\, dx = 1$, we have $\dfrac{6}{c} = 1$, so $c = 6$.

Solutions for Section 6.3

Exercises

1. $y = \displaystyle\int (x^3 + 5)\, dx = \dfrac{x^4}{4} + 5x + C$

5. Since $y = x + \sin x - \pi$, we differentiate to see that $dy/dx = 1 + \cos x$, so y satisfies the differential equation. To show that it also satisfies the initial condition, we check that $y(\pi) = 0$:

$$y = x + \sin x - \pi$$

$$y(\pi) = \pi + \sin \pi - \pi = 0.$$

9. Integrating gives

$$\int \frac{dq}{dz}\, dz = \int (2 + \sin z)\, dz = 2z - \cos z + C.$$

If $q = 5$ when $z = 0$, then $2(0) - \cos(0) + C = 5$ so $C = 6$. Thus $q = 2z - \cos z + 6$.

Problems

13.

$$\frac{dy}{dt} = k\sqrt{t} = kt^{1/2}$$

$$y = \frac{2}{3}kt^{3/2} + C.$$

Since $y = 0$ when $t = 0$, we have $C = 0$, so

$$y = \frac{2}{3}kt^{3/2}.$$

17. (a)

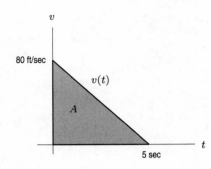

(b) The total distance is represented by the shaded region A, the area under the graph of $v(t)$.

(c) The area A, a triangle, is given by

$$A = \frac{1}{2}(\text{base})(\text{height}) = \frac{1}{2}(5 \text{ sec})(80 \text{ ft/sec}) = 200 \text{ ft}.$$

(d) Using integration and the Fundamental Theorem of Calculus, we have $A = \int_0^5 v(t)\, dt$ or $A = s(5) - s(0)$, where $s(t)$ is an antiderivative of $v(t)$.

We have that $a(t)$, the acceleration, is constant: $a(t) = k$ for some constant k. Therefore $v(t) = kt + C$ for some constant C. We have $80 = v(0) = k(0) + C = C$, so that $v(t) = kt + 80$. Putting in $t = 5$, $0 = v(5) = (k)(5) + 80$, or $k = -80/5 = -16$.

Thus $v(t) = -16t + 80$, and an antiderivative for $v(t)$ is $s(t) = -8t^2 + 80t + C$. Since the total distance traveled at $t = 0$ is 0, we have $s(0) = 0$ which means $C = 0$. Finally, $A = \int_0^5 v(t)\, dt = s(5) - s(0) = (-8(5)^2 + (80)(5)) - (-8(0)^2 + (80)(0)) = 200 \text{ ft}$, which agrees with the previous part.

21. The equation of motion is $y = -\frac{gt^2}{2} + v_0 t + y_0 = -16t^2 + 128t + 320$. Taking the first derivative, we get $v = -32t + 128$. The second derivative gives us $a = -32$.

(a) At its highest point, the stone's velocity is zero:
$v = 0 = -32t + 128$, so $t = 4$.

(b) At $t = 4$, the height is $y = -16(4)^2 + 128(4) + 320 = 576$ ft

(c) When the stone hits the beach,

$$y = 0 = -16t^2 + 128t + 320$$
$$0 = -t^2 + 8t + 20 = (10 - t)(2 + t).$$

So $t = 10$ seconds.

(d) Impact is at $t = 10$. The velocity, v, at this time is $v(10) = -32(10) + 128 = -192$ ft/sec. Upon impact, the stone's velocity is 192 ft/sec downward.

25. The first thing we should do is convert our units. We'll bring everything into feet and seconds. Thus, the initial speed of the car is

$$\frac{70 \text{ miles}}{\text{hour}} \left(\frac{1 \text{ hour}}{3600 \text{ sec}} \right) \left(\frac{5280 \text{ feet}}{1 \text{ mile}} \right) \approx 102.7 \text{ ft/sec}.$$

We assume that the acceleration is constant as the car comes to a stop. A graph of its velocity versus time is given in Figure 6.3. We know that the area under the curve represents the distance that the car travels before it comes to a stop, 157 feet. But this area is a triangle, so it is easy to find t_0, the time the car comes to rest. We solve

$$\frac{1}{2}(102.7)t_0 = 157,$$

which gives

$$t_0 \approx 3.06 \text{ sec}.$$

Since acceleration is the rate of change of velocity, the car's acceleration is given by the slope of the line in Figure 6.3. Thus, the acceleration, k, is given by

$$k = \frac{102.7 - 0}{0 - 3.06} \approx -33.56 \text{ ft/sec}^2.$$

Notice that k is negative because the car is slowing down.

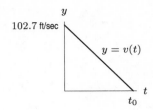

Figure 6.3: Graph of velocity versus time

Solutions for Section 6.4

Exercises

1.

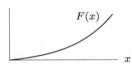

By the Fundamental Theorem, $f(x) = F'(x)$. Since f is positive and increasing, F is increasing and concave up. Since $F(0) = \int_0^0 f(t)dt = 0$, the graph of F must start from the origin.

5. Using the Fundamental Theorem, we know that the change in F between $x = 0$ and $x = 0.5$ is given by

$$F(0.5) - F(0) = \int_0^{0.5} \sin t \cos t \, dt \approx 0.115.$$

Since $F(0) = 1.0$, we have $F(0.5) \approx 1.115$. The other values are found similarly, and are given in Table 6.1.

Table 6.1

b	0	0.5	1	1.5	2	2.5	3
$F(b)$	1	1.11492	1.35404	1.4975	1.41341	1.17908	1.00996

9. If $f'(x) = \text{Si}(x)$, then $f(x)$ is of the form

$$f(x) = C + \int_a^x \text{Si}(t) \, dt.$$

Since $f(0) = 2$, we take $a = 0$ and $C = 2$, giving

$$f(x) = 2 + \int_0^x \text{Si}(t) \, dt.$$

Problems

13. Since $G'(x) = \cos(x^2)$ and $G(0) = -3$, we have

$$G(x) = G(0) + \int_0^x \cos(t^2) \, dt = -3 + \int_0^x \cos(t^2) \, dt.$$

Substituting $x = -1$ and evaluating the integral numerically gives

$$G(-1) = -3 + \int_0^{-1} \cos(t^2) \, dt = -3.905.$$

17. $\dfrac{d}{dt} \displaystyle\int_t^{\pi} \cos(z^3) \, dz = \dfrac{d}{dt} \left(-\int_{\pi}^t \cos(z^3) \, dz \right) = -\cos(t^3).$

21. (a) Since $\dfrac{d}{dt}(\cos(2t)) = -2\sin(2t)$, we have $F(\pi) = \displaystyle\int_0^\pi \sin(2t)\,dt = -\dfrac{1}{2}\cos(2t)\Big|_0^\pi = -\dfrac{1}{2}(1-1) = 0.$

(b) $F(\pi) = $ (Area above t-axis) $-$ (Area below t-axis) $= 0$. (The two areas are equal.)

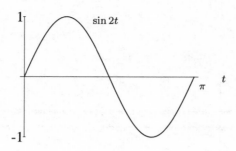

(c) $F(x) \geq 0$ everywhere. $F(x) = 0$ only at integer multiples of π. This can be seen for $x \geq 0$ by noting $F(x) = $ (Area above t-axis) $-$ (Area below t-axis), which is always non-negative and only equals zero when x is an integer multiple of π. For $x > 0$

$$F(-x) = \int_0^{-x} \sin 2t\,dt$$

$$= -\int_{-x}^0 \sin 2t\,dt$$

$$= \int_0^x \sin 2t\,dt = F(x),$$

since the area from $-x$ to 0 is the negative of the area from 0 to x. So we have $F(x) \geq 0$ for all x.

25. If we let $f(x) = \int_0^x e^{-t^2}\,dt$ and $g(x) = x^3$, then we use the chain rule because we are looking for $\dfrac{d}{dx}f(g(x)) = f'(g(x)) \cdot g'(x)$. Since $f'(x) = e^{-x^2}$, we have

$$\frac{d}{dx}\left(\int_0^{x^3} e^{-t^2}\,dt\right) = f'(x^3) \cdot 3x^2 = e^{-(x^3)^2} \cdot 3x^2 = 3x^2 e^{-x^6}.$$

Solutions for Section 6.5

Exercises

1. (a) The object is thrown from an initial height of $y = 1.5$ meters.
(b) The velocity is obtained by differentiating, which gives $v = -9.8t + 7$ m/sec. The initial velocity is $v = 7$ m/sec upward.
(c) The acceleration due to gravity is obtained by differentiating again, giving $g = -9.8$ m/sec^2, or 9.8 m/sec^2 downward.

Problems

5. $a(t) = -32$. Since $v(t)$ is the antiderivative of $a(t)$, $v(t) = -32t + v_0$. But $v_0 = 0$, so $v(t) = -32t$. Since $s(t)$ is the antiderivative of $v(t)$, $s(t) = -16t^2 + s_0$, where s_0 is the height of the building. Since the ball hits the ground in 5 seconds, $s(5) = 0 = -400 + s_0$. Hence $s_0 = 400$ feet, so the window is 400 feet high.

9. (a) Since $s(t) = -\frac{1}{2}gt^2$, the distance a body falls in the first second is

$$s(1) = -\frac{1}{2} \cdot g \cdot 1^2 = -\frac{g}{2}.$$

In the second second, the body travels

$$s(2) - s(1) = -\frac{1}{2}\left(g \cdot 2^2 - g \cdot 1^2\right) = -\frac{1}{2}(4g - g) = -\frac{3g}{2}.$$

In the third second, the body travels

$$s(3) - s(2) = -\frac{1}{2}\left(g \cdot 3^2 - g \cdot 2^2\right) = -\frac{1}{2}(9g - 4g) = -\frac{5g}{2},$$

and in the fourth second, the body travels

$$s(4) - s(3) = -\frac{1}{2}\left(g \cdot 4^2 - g \cdot 3^2\right) = -\frac{1}{2}(16g - 9g) = -\frac{7g}{2}.$$

   (b) Galileo seems to have been correct. His observation follows from the fact that the differences between consecutive squares are consecutive odd numbers. For, if n is any number, then $n^2 - (n-1)^2 = 2n - 1$, which is the nth odd number (where 1 is the first).

Solutions for Chapter 6 Review

Exercises

1. $\frac{5}{2}x^2 + 7x + C$

5. $\displaystyle\int (3e^x + 2\sin x)\, dx = 3e^x - 2\cos x + C$

9. $e^x + 5x + C$

13. $\displaystyle\int (x+1)^2\, dx = \frac{(x+1)^3}{3} + C.$

    Another way to work the problem is to expand $(x+1)^2$ to $x^2 + 2x + 1$ as follows:

    $$\int (x+1)^2\, dx = \int (x^2 + 2x + 1)\, dx = \frac{x^3}{3} + x^2 + x + C.$$

    These two answers are the same, since $\dfrac{(x+1)^3}{3} = \dfrac{x^3 + 3x^2 + 3x + 1}{3} = \dfrac{x^3}{3} + x^2 + x + \dfrac{1}{3}$, which is $\dfrac{x^3}{3} + x^2 + x$, plus a constant.

17. Since $f(x) = x + 1 + \dfrac{1}{x}$, the indefinite integral is $\dfrac{1}{2}x^2 + x + \ln|x| + C$

21. $2e^x - 8\sin x + C$

25. $G(x) = \displaystyle\int \sin x\, dx = -\cos x + C$

29. $F(x) = \displaystyle\int (e^x - 1)\, dx = e^x - x + C$

33. $F(x) = \displaystyle\int e^x\, dx = e^x + C.$ If $F(0) = 4$, then $F(0) = 1 + C = 4$ and thus $C = 3$. So $F(x) = e^x + 3.$

Problems

37. $\displaystyle\int_0^3 x^2\, dx = \frac{x^3}{3}\bigg|_0^3 = 9 - 0 = 9.$

41. (a) See Figure 6.4. Since $f(x) > 0$ for $0 < x < 2$ and $f(x) < 0$ for $2 < x < 5$, we have

    $$\text{Area} = \int_0^2 f(x)\, dx - \int_2^5 f(x)\, dx$$

    $$= \int_0^2 (x^3 - 7x^2 + 10x)\, dx - \int_2^5 (x^3 - 7x^2 + 10x)\, dx$$

$$= \left(\frac{x^4}{4} - \frac{7x^3}{3} + 5x^2 \right) \Big|_0^2 - \left(\frac{x^4}{4} - \frac{7x^3}{3} + 5x^2 \right) \Big|_2^5$$

$$= \left[\left(4 - \frac{56}{3} + 20 \right) - (0 - 0 + 0) \right] - \left[\left(\frac{625}{4} - \frac{875}{3} + 125 \right) - \left(4 - \frac{56}{3} + 20 \right) \right]$$

$$= \frac{253}{12}.$$

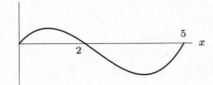

Figure 6.4: Graph of $f(x) = x^3 - 7x^2 + 10x$

(b) Calculating $\int_0^5 f(x)\, dx$ gives

$$\int_0^5 f(x)\, dx = \int_0^5 (x^3 - 7x^2 + 10x)\, dx$$

$$= \left(\frac{x^4}{4} - \frac{7x^3}{3} + 5x^2 \right) \Big|_0^5$$

$$= \left(\frac{625}{4} - \frac{875}{3} + 125 \right) - (0 - 0 + 0)$$

$$= -\frac{125}{12}.$$

This integral measures the difference between the area above the x-axis and the area below the x-axis. Since the definite integral is negative, the graph of $f(x)$ lies more below the x-axis than above it. Since the function crosses the axis at $x = 2$,

$$\int_0^5 f(x)\, dx = \int_0^2 f(x)\, dx + \int_2^5 f(x)\, dx = \frac{16}{3} - \frac{63}{4} = \frac{-125}{12},$$

whereas

$$\text{Area} = \int_0^2 f(x)\, dx - \int_2^5 f(x)\, dx = \frac{16}{3} + \frac{64}{4} = \frac{253}{12}.$$

45. See Figure 6.5. The average value of $f(x)$ is given by

$$\text{Average} = \frac{1}{9 - 0} \int_0^9 \sqrt{x}\, dx = \frac{1}{9} \left(\frac{2}{3} x^{3/2} \Big|_0^9 \right) = \frac{1}{9} \left(\frac{2}{3} 9^{3/2} - 0 \right) = \frac{1}{9} 18 = 2.$$

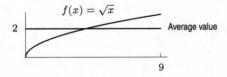

Figure 6.5

49.

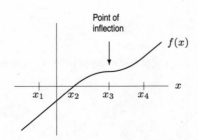

53. A function whose derivative is e^{x^2} is of the form

$$f(x) = C + \int_a^x e^{t^2}\, dt \qquad \text{for some value of } C.$$

(a) To ensure that the function goes through the point $(0, 3)$, we take $a = 0$ and $C = 3$:

$$f(x) = 3 + \int_0^x e^{t^2}\, dt.$$

(b) To ensure that the function goes through $(-1, 5)$, we take $a = -1$ and $C = 5$:

$$f(x) = 5 + \int_{-1}^x e^{t^2}\, dt.$$

57. (a) Since $6 \text{ sec} = 1/10$ min,

$$\text{Angular acceleration} = \frac{2500 - 1100}{1/10} = 14{,}000 \text{ revs/min}^2.$$

(b) We know angular acceleration is the derivative of angular velocity. Since

$$\text{Angular acceleration} = 14{,}000,$$

we have

$$\text{Angular velocity} = 14{,}000t + C.$$

Measuring time from the moment at which the angular velocity is 1100 revs/min, we have $C = 1100$. Thus,

$$\text{Angular velocity} = 14{,}000t + 1100.$$

Thus the total number of revolutions performed during the period from $t = 0$ to $t = 1/10$ min is given by

$$\begin{array}{l} \text{Number of} \\ \text{revolutions} \end{array} = \int_0^{1/10} (14000t + 1100)dt = 7000t^2 + 1100t \Big|_0^{1/10} = 180 \text{ revolutions.}$$

61. (a) In the beginning, both birth and death rates are small; this is consistent with a very small population. Both rates begin climbing, the birth rate faster than the death rate, which is consistent with a growing population. The birth rate is then high, but it begins to decrease as the population increases.

(b)

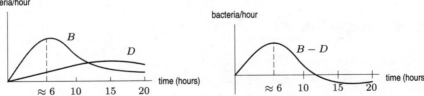

Figure 6.6: Difference between B and D is greatest at $t \approx 6$

The bacteria population is growing most quickly when $B - D$, the rate of change of population, is maximal; that happens when B is farthest above D, which is at a point where the slopes of both graphs are equal. That point is $t \approx 6$ hours.

(c) Total number born by time t is the area under the B graph from $t = 0$ up to time t. See Figure 6.7.

Total number alive at time t is the number born minus the number that have died, which is the area under the B graph minus the area under the D graph, up to time t. See Figure 6.8.

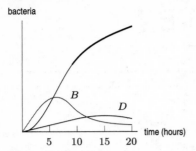

Figure 6.7: Number born by time t is
$$\int_0^t B(x)\,dx$$

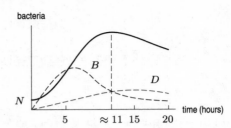

Figure 6.8: Number alive at time t is
$$\int_0^t (B(x) - D(x))\,dx$$

From Figure 6.8, we see that the population is at a maximum when $B = D$, that is, after about 11 hours. This stands to reason, because $B - D$ is the rate of change of population, so population is maximized when $B - D = 0$, that is, when $B = D$.

CAS Challenge Problems

65. (a) A CAS gives
$$\int e^{2x}\,dx = \frac{1}{2}e^{2x} \qquad \int e^{3x}\,dx = \frac{1}{3}e^{3x} \qquad \int e^{3x+5}\,dx = \frac{1}{3}e^{3x+5}.$$

(b) The three integrals in part (a) obey the rule
$$\int e^{ax+b}\,dx = \frac{1}{a}e^{ax+b}.$$

(c) Checking the formula by calculating the derivative
$$\frac{d}{dx}\left(\frac{1}{a}e^{ax+b}\right) = \frac{1}{a}\frac{d}{dx}e^{ax+b} \quad \text{by the constant multiple rule}$$
$$= \frac{1}{a}e^{ax+b}\frac{d}{dx}(ax+b) \quad \text{by the chain rule}$$
$$= \frac{1}{a}e^{ax+b} \cdot a = e^{ax+b}.$$

CHECK YOUR UNDERSTANDING

1. True. A function can have only one derivative.

5. False. Differentiating using the product and chain rules gives
$$\frac{d}{dx}\left(\frac{-1}{2x}e^{-x^2}\right) = \frac{1}{2x^2}e^{-x^2} + e^{-x^2}.$$

9. True. If $y = F(x)$ is a solution to the differential equation $dy/dx = f(x)$, then $F'(x) = f(x)$, so $F(x)$ is an antiderivative of $f(x)$.

13. False. If f is positive then F is increasing, but if f is negative then F is decreasing.

CHAPTER SEVEN

Solutions for Section 7.1

Exercises

1. (a) $\frac{d}{dx}\sin(x^2+1) = 2x\cos(x^2+1);$ $\qquad$ $\frac{d}{dx}\sin(x^3+1) = 3x^2\cos(x^3+1)$
   (b) (i) $\frac{1}{2}\sin(x^2+1) + C$ $\qquad$ (ii) $\frac{1}{3}\sin(x^3+1) + C$
   (c) (i) $-\frac{1}{2}\cos(x^2+1) + C$ $\qquad$ (ii) $-\frac{1}{3}\cos(x^3+1) + C$

5. We use the substitution $w = 2x$, $dw = 2\,dx$.

$$\int \sin(2x)\,dx = \frac{1}{2}\int \sin(w)\,dw = -\frac{1}{2}\cos(w) + C = -\frac{1}{2}\cos(2x) + C.$$

Check: $\frac{d}{dx}\left(-\frac{1}{2}\cos(2x) + C\right) = \frac{1}{2}\sin(2x)(2) = \sin(2x)$.

9. In this case, it seems easier not to substitute.

$$\int y^2(1+y)^2\,dy = \int y^2(y^2+2y+1)\,dy = \int (y^4 + 2y^3 + y^2)\,dy$$
$$= \frac{y^5}{5} + \frac{y^4}{2} + \frac{y^3}{3} + C.$$

Check: $\frac{d}{dy}\left(\frac{y^5}{5} + \frac{y^4}{2} + \frac{y^3}{3} + C\right) = y^4 + 2y^3 + y^2 = y^2(y+1)^2$.

13. We use the substitution $w = 4 - x$, $dw = -dx$.

$$\int \frac{1}{\sqrt{4-x}}\,dx = -\int \frac{1}{\sqrt{w}}\,dw = -2\sqrt{w} + C = -2\sqrt{4-x} + C.$$

Check: $\frac{d}{dx}(-2\sqrt{4-x} + C) = -2 \cdot \frac{1}{2} \cdot \frac{1}{\sqrt{4-x}} \cdot -1 = \frac{1}{\sqrt{4-x}}$.

17. We use the substitution $w = \cos\theta + 5$, $dw = -\sin\theta\,d\theta$.

$$\int \sin\theta(\cos\theta + 5)^7\,d\theta = -\int w^7\,dw = -\frac{1}{8}w^8 + C$$
$$= -\frac{1}{8}(\cos\theta + 5)^8 + C.$$

Check:

$$\frac{d}{d\theta}\left[-\frac{1}{8}(\cos\theta + 5)^8 + C\right] = -\frac{1}{8} \cdot 8(\cos\theta + 5)^7 \cdot (-\sin\theta)$$
$$= \sin\theta(\cos\theta + 5)^7$$

21. We use the substitution $w = x^3 + 1$, $dw = 3x^2\,dx$, to get

$$\int x^2 e^{x^3+1}\,dx = \frac{1}{3}\int e^w\,dw = \frac{1}{3}e^w + C = \frac{1}{3}e^{x^3+1} + C.$$

Check: $\frac{d}{dx}\left(\frac{1}{3}e^{x^3+1} + C\right) = \frac{1}{3}e^{x^3+1} \cdot 3x^2 = x^2 e^{x^3+1}$.

25. We use the substitution $w = \ln z$, $dw = \frac{1}{z}\,dz$.

$$\int \frac{(\ln z)^2}{z}\,dz = \int w^2\,dw = \frac{w^3}{3} + C = \frac{(\ln z)^3}{3} + C.$$

Check: $\frac{d}{dz}\left[\frac{(\ln z)^3}{3} + C\right] = 3 \cdot \frac{1}{3}(\ln z)^2 \cdot \frac{1}{z} = \frac{(\ln z)^2}{z}$.

29. We use the substitution $w = \sqrt{y}$, $dw = \dfrac{1}{2\sqrt{y}} \, dy$.

$$\int \frac{e^{\sqrt{y}}}{\sqrt{y}} \, dy = 2 \int e^w \, dw = 2e^w + C = 2e^{\sqrt{y}} + C.$$

Check: $\dfrac{d}{dy}(2e^{\sqrt{y}} + C) = 2e^{\sqrt{y}} \cdot \dfrac{1}{2\sqrt{y}} = \dfrac{e^{\sqrt{y}}}{\sqrt{y}}.$

33. We use the substitution $w = 1 + 3t^2$, $dw = 6t \, dt$.

$$\int \frac{t}{1 + 3t^2} \, dt = \int \frac{1}{w}\left(\frac{1}{6} \, dw\right) = \frac{1}{6} \ln|w| + C = \frac{1}{6} \ln(1 + 3t^2) + C.$$

(We can drop the absolute value signs since $1 + 3t^2 > 0$ for all t).

Check: $\dfrac{d}{dt}\left[\dfrac{1}{6} \ln(1 + 3t^2) + C\right] = \dfrac{1}{6}\dfrac{1}{1 + 3t^2}(6t) = \dfrac{t}{1 + 3t^2}.$

37. Since $d(\sinh x)/dx = \cosh x$, we have

$$\int \cosh x \, dx = \sinh x + C.$$

41. The general antiderivative is $\int (\pi t^3 + 4t) \, dt = (\pi/4)t^4 + 2t^2 + C.$

45. Make the substitution $w = 2 - 5x$, then $dw = -5dx$. We have

$$\int \sin(2 - 5x)dx = \int \sin w \left(-\frac{1}{5}\right) dw = -\frac{1}{5}(-\cos w) + C = \frac{1}{5}\cos(2 - 5x) + C.$$

49. $\displaystyle\int_0^\pi \cos(x + \pi) \, dx = \sin(x + \pi)\Big|_0^\pi = \sin(2\pi) - \sin(\pi) = 0 - 0 = 0$

53. We substitute $w = \sqrt[3]{x} = x^{\frac{1}{3}}$. Then $dw = \dfrac{1}{3}x^{-\frac{2}{3}} \, dx = \dfrac{1}{3\sqrt[3]{x^2}} \, dx.$

$$\int_1^8 \frac{e^{\sqrt[3]{x}}}{\sqrt[3]{x^2}}dx = \int_{x=1}^{x=8} e^w (3 \, dw) = 3e^w \Big|_{x=1}^{x=8} = 3e^{\sqrt[3]{x}}\Big|_1^8 = 3(e^2 - e).$$

57.
$$\int_{-1}^3 (x^3 + 5x) \, dx = \frac{x^4}{4}\bigg|_{-1}^3 + \frac{5x^2}{2}\bigg|_{-1}^3 = 40.$$

61. $\displaystyle\int_{-1}^2 \sqrt{x + 2} \, dx = \frac{2}{3}(x + 2)^{3/2}\bigg|_{-1}^2 = \frac{2}{3}\left[(4)^{3/2} - (1)^{3/2}\right] = \frac{2}{3}(7) = \frac{14}{3}$

Problems

65. Since $f(x) = 1/(x + 1)$ is positive on the interval $x = 0$ to $x = 2$, we have

$$\text{Area} = \int_0^2 \frac{1}{x + 1}dx = \ln(x + 1)\bigg|_0^2 = \ln 3 - \ln 1 = \ln 3.$$

The area is $\ln 3 \approx 1.0986$.

69. If $f(x) = \dfrac{1}{x + 1}$, the average value of f on the interval $0 \le x \le 2$ is defined to be

$$\frac{1}{2 - 0}\int_0^2 f(x) \, dx = \frac{1}{2}\int_0^2 \frac{dx}{x + 1}.$$

We'll integrate by substitution. We let $w = x + 1$ and $dw = dx$, and we have

$$\int_{x=0}^{x=2} \frac{dx}{x + 1} = \int_{w=1}^{w=3} \frac{dw}{w} = \ln w\bigg|_1^3 = \ln 3 - \ln 1 = \ln 3.$$

Thus, the average value of $f(x)$ on $0 \le x \le 2$ is $\frac{1}{2}\ln 3 \approx 0.5493$. See Figure 7.1.

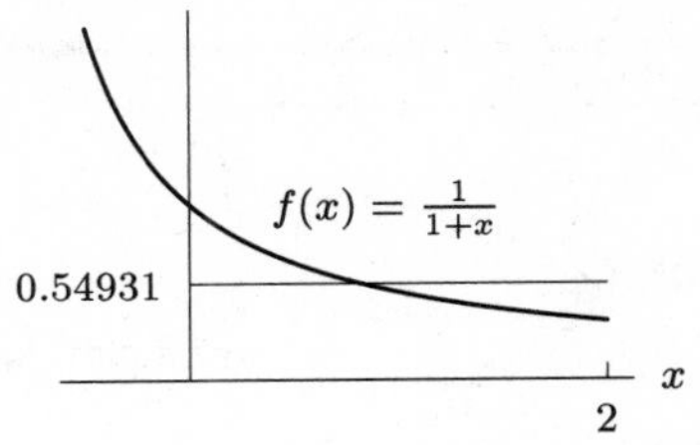

Figure 7.1

73. **(a)** In 1990, we have $P = 5.3e^{0.014(0)} = 5.3$ billion people.
    In 2000, we have $P = 5.3e^{0.014(10)} = 6.1$ billion people.
    (b) We have

$$\text{Average population} = \frac{1}{10-0} \int_0^{10} 5.3 e^{0.014t}\, dt = \frac{1}{10} \cdot \frac{5.3}{0.014} e^{0.014t} \Big|_0^{10}$$

$$= \frac{1}{10} \left(\frac{5.3}{0.014} (e^{0.14} - e^0) \right) = 5.7.$$

The average population of the world during the 1990s was 5.7 billion people.

77. Since v is given as the velocity of a falling body, the height h is decreasing, so $v = -\frac{dh}{dt}$, and it follows that $h(t) = -\int v(t)\, dt$ and $h(0) = h_0$. Let $w = e^{t\sqrt{gk}} + e^{-t\sqrt{gk}}$. Then

$$dw = \sqrt{gk} \left(e^{t\sqrt{gk}} - e^{-t\sqrt{gk}} \right) dt,$$

so $\dfrac{dw}{\sqrt{gk}} = (e^{t\sqrt{gk}} - e^{-t\sqrt{gk}})\, dt$. Therefore,

$$-\int v(t)dt = -\int \sqrt{\frac{g}{k}} \left(\frac{e^{t\sqrt{gk}} - e^{-t\sqrt{gk}}}{e^{t\sqrt{gk}} + e^{-t\sqrt{gk}}} \right) dt$$

$$= -\sqrt{\frac{g}{k}} \int \frac{1}{e^{t\sqrt{gk}} + e^{-t\sqrt{gk}}} \left(e^{t\sqrt{gk}} - e^{-t\sqrt{gk}} \right) dt$$

$$= -\sqrt{\frac{g}{k}} \int \left(\frac{1}{w} \right) \frac{dw}{\sqrt{gk}}$$

$$= -\sqrt{\frac{g}{gk^2}} \ln |w| + C$$

$$= -\frac{1}{k} \ln \left(e^{t\sqrt{gk}} + e^{-t\sqrt{gk}} \right) + C.$$

Since

$$h(0) = -\frac{1}{k} \ln(e^0 + e^0) + C = -\frac{\ln 2}{k} + C = h_0,$$

we have $C = h_0 + \dfrac{\ln 2}{k}$. Thus,

$$h(t) = -\frac{1}{k} \ln \left(e^{t\sqrt{gk}} + e^{-t\sqrt{gk}} \right) + \frac{\ln 2}{k} + h_0 = -\frac{1}{k} \ln \left(\frac{e^{t\sqrt{gk}} + e^{-t\sqrt{gk}}}{2} \right) + h_0.$$

Solutions for Section 7.2

Exercises

1. Let $u = \arctan x$, $v' = 1$. Then $v = x$ and $u' = \dfrac{1}{1+x^2}$. Integrating by parts, we get:

$$\int 1 \cdot \arctan x \, dx = x \cdot \arctan x - \int x \cdot \frac{1}{1+x^2} \, dx.$$

To compute the second integral use the substitution, $z = 1 + x^2$.

$$\int \frac{x}{1+x^2} \, dx = \frac{1}{2} \int \frac{dz}{z} = \frac{1}{2} \ln|z| + C = \frac{1}{2} \ln(1+x^2) + C.$$

Thus,

$$\int \arctan x \, dx = x \cdot \arctan x - \frac{1}{2} \ln(1+x^2) + C.$$

5. Let $u = t$, $v' = \sin t$. Thus, $v = -\cos t$ and $u' = 1$. With this choice of u and v, integration by parts gives:

$$\int t \sin t \, dt = -t \cos t - \int (-\cos t) \, dt$$
$$= -t \cos t + \sin t + C.$$

9. Let $u = t^2$, $v' = \sin t$ implying $v = -\cos t$ and $u' = 2t$. Integrating by parts, we get:

$$\int t^2 \sin t \, dt = -t^2 \cos t - \int 2t(-\cos t) \, dt.$$

Again, applying integration by parts with $u = t$, $v' = \cos t$, we have:

$$\int t \cos t \, dt = t \sin t + \cos t + C.$$

Thus

$$\int t^2 \sin t \, dt = -t^2 \cos t + 2t \sin t + 2 \cos t + C.$$

13. Let $u = \ln 5q$, $v' = q^5$. Then $v = \frac{1}{6} q^6$ and $u' = \dfrac{1}{q}$. Integrating by parts, we get:

$$\int q^5 \ln 5q \, dq = \frac{1}{6} q^6 \ln 5q - \int \left(5 \cdot \frac{1}{5q}\right) \cdot \frac{1}{6} q^6 \, dq$$
$$= \frac{1}{6} q^6 \ln 5q - \frac{1}{36} q^6 + C.$$

17. Let $u = \theta + 1$ and $v' = \sin(\theta + 1)$, so $u' = 1$ and $v = -\cos(\theta + 1)$.

$$\int (\theta + 1) \sin(\theta + 1) \, d\theta = -(\theta + 1) \cos(\theta + 1) + \int \cos(\theta + 1) \, d\theta$$
$$= -(\theta + 1) \cos(\theta + 1) + \sin(\theta + 1) + C.$$

21. $\displaystyle \int \frac{t+7}{\sqrt{5-t}} \, dt = \int \frac{t}{\sqrt{5-t}} \, dt + 7 \int (5-t)^{-1/2} \, dt.$

To calculate the first integral, we use integration by parts. Let $u = t$ and $v' = \frac{1}{\sqrt{5-t}}$, so $u' = 1$ and $v = -2(5-t)^{1/2}$. Then

$$\int \frac{t}{\sqrt{5-t}} \, dt = -2t(5-t)^{1/2} + 2 \int (5-t)^{1/2} \, dt = -2t(5-t)^{1/2} - \frac{4}{3}(5-t)^{3/2} + C.$$

We can calculate the second integral directly: $7 \displaystyle\int (5-t)^{-1/2} = -14(5-t)^{1/2} + C_1$. Thus

$$\int \frac{t+7}{\sqrt{5-t}} \, dt = -2t(5-t)^{1/2} - \frac{4}{3}(5-t)^{3/2} - 14(5-t)^{1/2} + C_2.$$

25. This integral can first be simplified by making the substitution $w = x^2$, $dw = 2x\,dx$. Then

$$\int x \arctan x^2 \, dx = \frac{1}{2} \int \arctan w \, dw.$$

To evaluate $\int \arctan w \, dw$, we'll use integration by parts. Let $u = \arctan w$ and $v' = 1$, so $u' = \frac{1}{1+w^2}$ and $v = w$. Then

$$\int \arctan w \, dw = w \arctan w - \int \frac{w}{1+w^2} \, dw = w \arctan w - \frac{1}{2} \ln|1 + w^2| + C.$$

Since $1 + w^2$ is never negative, we can drop the absolute value signs. Thus, we have

$$\int x \arctan x^2 \, dx = \frac{1}{2}\left(x^2 \arctan x^2 - \frac{1}{2} \ln(1 + (x^2)^2) + C \right)$$

$$= \frac{1}{2} x^2 \arctan x^2 - \frac{1}{4} \ln(1 + x^4) + C.$$

29. $\displaystyle \int_3^5 x \cos x \, dx = (\cos x + x \sin x)\Big|_3^5 = \cos 5 + 5 \sin 5 - \cos 3 - 3 \sin 3 \approx -3.944.$

33. $\displaystyle \int_0^5 \ln(1 + t) \, dt = ((1 + t)\ln(1 + t) - (1 + t))\Big|_0^5 = 6 \ln 6 - 5 \approx 5.751.$

Problems

37.

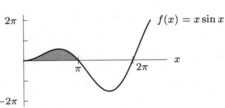

The graph of $f(x) = x \sin x$ is shown above. The first positive zero is at $x = \pi$, so, using integration by parts,

$$\text{Area} = \int_0^\pi x \sin x \, dx$$

$$= -x \cos x \Big|_0^\pi + \int_0^\pi \cos x \, dx$$

$$= -x \cos x \Big|_0^\pi + \sin x \Big|_0^\pi$$

$$= -\pi \cos \pi - (-0 \cos 0) + \sin \pi - \sin 0 = \pi.$$

41. Let $u = e^\theta$ and $v' = \cos \theta$, so $u' = e^\theta$ and $v = \sin \theta$. Then $\int e^\theta \cos \theta \, d\theta = e^\theta \sin \theta - \int e^\theta \sin \theta \, d\theta$.

In Problem 40 we found that $\int e^x \sin x \, dx = \frac{1}{2} e^x (\sin x - \cos x) + C$.

$$\int e^\theta \cos \theta \, d\theta = e^\theta \sin \theta - \left[\frac{1}{2} e^\theta (\sin \theta - \cos \theta) \right] + C$$

$$= \frac{1}{2} e^\theta (\sin \theta + \cos \theta) + C.$$

45. We integrate by parts. Let $u = x^n$ and $v' = \cos ax$, so $u' = nx^{n-1}$ and $v = \frac{1}{a} \sin ax$. Then

$$\int x^n \cos ax \, dx = \frac{1}{a} x^n \sin ax - \int (nx^{n-1})(\frac{1}{a} \sin ax) \, dx$$

$$= \frac{1}{a} x^n \sin ax - \frac{n}{a} \int x^{n-1} \sin ax \, dx.$$

49. Since $f'(x) = 2x$, integration by parts tells us that

$$\int_0^{10} f(x)g'(x)\,dx = f(x)g(x)\Big|_0^{10} - \int_0^{10} f'(x)g(x)\,dx$$

$$= f(10)g(10) - f(0)g(0) - 2\int_0^{10} xg(x)\,dx.$$

We can use left and right Riemann Sums with $\Delta x = 2$ to approximate $\int_0^{10} xg(x)\,dx$:

$$\text{Left sum} \approx 0 \cdot g(0)\Delta x + 2 \cdot g(2)\Delta x + 4 \cdot g(4)\Delta x + 6 \cdot g(6)\Delta x + 8 \cdot g(8)\Delta x$$
$$= (0(2.3) + 2(3.1) + 4(4.1) + 6(5.5) + 8(5.9))\,2 = 205.6.$$
$$\text{Right sum} \approx 2 \cdot g(2)\Delta x + 4 \cdot g(4)\Delta x + 6 \cdot g(6)\Delta x + 8 \cdot g(8)\Delta x + 10 \cdot g(10)\Delta x$$
$$= (2(3.1) + 4(4.1) + 6(5.5) + 8(5.9) + 10(6.1))\,2 = 327.6.$$

A good estimate for the integral is the average of the left and right sums, so

$$\int_0^{10} xg(x)\,dx \approx \frac{205.6 + 327.6}{2} = 266.6.$$

Substituting values for f and g, we have

$$\int_0^{10} f(x)g'(x)\,dx = f(10)g(10) - f(0)g(0) - 2\int_0^{10} xg(x)\,dx$$

$$\approx 10^2(6.1) - 0^2(2.3) - 2(266.6) = 76.8 \approx 77.$$

53. (a) Increasing V_0 increases the maximum value of V, since this maximum is V_0. Increasing ω or ϕ does not affect the maximum of V.

   (b) Since

   $$\frac{dV}{dt} = -\omega V_0 \sin(\omega t + \phi),$$

   the maximum of dV/dt is ωV_0. Thus, the maximum of dV/dt is increased if V_0 or ω is increased, and is unaffected if ϕ is increased.

   (c) The period of $V = V_0 \cos(\omega t + \phi)$ is $2\pi/\omega$, so

   $$\text{Average value} = \frac{1}{2\pi/\omega} \int_0^{2\pi/\omega} (V_0 \cos(\omega t + \phi))^2\,dt.$$

   Substituting $x = \omega t + \phi$, we have $dx = \omega dt$. When $t = 0$, $x = \phi$, and when $t = 2\pi/\omega$, $x = 2\pi + \phi$. Thus,

   $$\text{Average value} = \frac{\omega}{2\pi} \int_\phi^{2\pi+\phi} V_0^2 (\cos x)^2 \frac{1}{\omega}\,dx$$

   $$= \frac{V_0^2}{2\pi} \int_\phi^{2\pi+\phi} (\cos x)^2\,dx.$$

   Using integration by parts and the fact that $\sin^2 x = 1 - \cos^2 x$, we see that

   $$\text{Average value} = \frac{V_0^2}{2\pi} \left[\frac{1}{2}(\cos x \sin x + x)\right]_\phi^{2\pi+\phi}$$

   $$= \frac{V_0^2}{4\pi} [\cos(2\pi + \phi)\sin(2\pi + \phi) + (2\pi + \phi) - \cos\phi\sin\phi - \phi]$$

   $$= \frac{V_0^2}{4\pi} \cdot 2\pi = \frac{V_0^2}{2}.$$

   Thus, increasing V_0 increases the average value; increasing ω or ϕ has no effect.
   However, it is not in fact necessary to compute the integral to see that ω does not affect the average value, since all ω's dropped out of the average value expression when we made the substitution $x = \omega t + \phi$.

Solutions for Section 7.3

Exercises

1. $\frac{1}{10}e^{(-3\theta)}(-3\cos\theta + \sin\theta) + C.$
   (Let $a = -3, b = 1$ in II-9.)

5. Note that you can't use substitution here: letting $w = x^3 + 5$ doesn't work, since there is no $dw = 3x^2\,dx$ in the integrand. What will work is simply multiplying out the square: $(x^3 + 5)^2 = x^6 + 10x^3 + 25$. Then use I-1:

$$\int (x^3 + 5)^2\,dx = \int x^6\,dx + 10\int x^3\,dx + 25\int 1\,dx = \frac{1}{7}x^7 + 10\cdot\frac{1}{4}x^4 + 25x + C.$$

9. Let $m = 3$ in IV-21.

$$\int \frac{1}{\cos^3 x}\,dx = \frac{1}{2}\frac{\sin x}{\cos^2 x} + \frac{1}{2}\int \frac{1}{\cos x}\,dx$$

$$= \frac{1}{2}\frac{\sin x}{\cos^2 x} + \frac{1}{4}\ln\left|\frac{\sin x + 1}{\sin x - 1}\right| + C \text{ by IV-22.}$$

13. $\left(\frac{1}{3}x^2 - \frac{2}{9}x + \frac{2}{27}\right)e^{3x} + C.$
    (Let $a = 3, p(x) = x^2$ in III-14.)

17. Use long division to reorganize the integral:

$$\int \frac{t^2 + 1}{t^2 - 1}\,dt = \int \left(1 + \frac{2}{t^2 - 1}\right)\,dt = \int dt + \int \frac{2}{(t-1)(t+1)}\,dt.$$

To get this second integral, let $a = 1, b = -1$ in V-26, so

$$\int \frac{t^2 + 1}{t^2 - 1}\,dt = t + \ln|t - 1| - \ln|t + 1| + C.$$

21. $\frac{1}{34}e^{5x}(5\sin 3x - 3\cos 3x) + C.$
    (Let $a = 5, b = 3$ in II-8.)

25. Substitute $w = 7x, dw = 7\,dx$. Then use IV-21.

$$\int \frac{1}{\cos^4 7x}\,dx = \frac{1}{7}\int \frac{1}{\cos^4 w}\,dw = \frac{1}{7}\left[\frac{1}{3}\frac{\sin w}{\cos^3 w} + \frac{2}{3}\int \frac{1}{\cos^2 w}\,dw\right]$$

$$= \frac{1}{21}\frac{\sin w}{\cos^3 w} + \frac{2}{21}\left[\frac{\sin w}{\cos w} + C\right]$$

$$= \frac{1}{21}\frac{\tan w}{\cos^2 w} + \frac{2}{21}\tan w + C$$

$$= \frac{1}{21}\frac{\tan 7x}{\cos^2 7x} + \frac{2}{21}\tan 7x + C.$$

29.

$$\int \frac{dy}{4 - y^2} = -\int \frac{dy}{(y+2)(y-2)} = -\frac{1}{4}(\ln|y - 2| - \ln|y + 2|) + C.$$

(Let $a = 2, b = -2$ in V-26.)

33. We use the Pythagorean Identity to change the integrand in the following manner:

$$\sin^3 x = \left(\sin^2 x\right)\sin x = \left(1 - \cos^2 x\right)\sin x = \sin x - \cos^2 x\sin x.$$

Thus, we have

$$\int \sin^3 x \, dx = \int \left(\sin x - \cos^2 x \sin x\right) dx$$

$$= \int \sin x \, dx - \int \cos^2 x \sin x \, dx.$$

The first of these new integrals can be easily found. The second can be found using the substitution $w = \cos x$ so $dw = -\sin x \, dx$. The second integral becomes

$$\int \cos^2 x \sin x \, dx = -\int w^2 dw$$

$$= -\frac{1}{3}w^3 + C$$

$$= -\frac{1}{3}\cos^3 x + C$$

and so our final answer is

$$\int \sin^3 x \, dx = \int \sin x \, dx - \int \cos^2 x \sin x \, dx$$

$$= -\cos x + (1/3)\cos^3 x + C.$$

Problems

37. Using formula II-11, if $m \neq \pm n$, then

$$\int_{-\pi}^{\pi} \cos m\theta \cos n\theta \, d\theta = \frac{1}{n^2 - m^2}(n \cos m\theta \sin n\theta - m \sin m\theta \cos n\theta)\Big|_{-\pi}^{\pi}.$$

We see that in the evaluation, each term will have a $\sin k\pi$ term, so the expression reduces to 0.

Solutions for Section 7.4

Exercises

1. Since $25 - x^2 = (5 - x)(5 + x)$, we take

$$\frac{20}{25 - x^2} = \frac{A}{5 - x} + \frac{B}{5 + x}.$$

So,

$$20 = A(5 + x) + B(5 - x)$$
$$20 = (A - B)x + 5A + 5B,$$

giving

$$A - B = 0$$
$$5A + 5B = 20.$$

Thus $A = B = 2$ and

$$\frac{20}{25 - x^2} = \frac{2}{5 - x} + \frac{2}{5 + x}.$$

5. Using the result of Problem 1, we have

$$\int \frac{20}{25 - x^2} \, dx = \int \frac{2}{5 - x} \, dx + \int \frac{2}{5 + x} \, dx = -2\ln|5 - x| + 2\ln|5 + x| + C.$$

9. (a) Yes, use $x = 3\sin\theta$.
   (b) No; better to substitute $w = 9 - x^2$, so $dw = -2x \, dx$.

13. Division gives

$$\frac{x^4 + 12x^3 + 15x^2 + 25x + 11}{x^3 + 12x^2 + 11x} = x + \frac{4x^2 + 25x + 11}{x^3 + 12x^2 + 11x}.$$

Since $x^3 + 12x^2 + 11x = x(x+1)(x+11)$, we write

$$\frac{4x^2 + 25x + 11}{x^3 + 12x^2 + 11x} = \frac{A}{x} + \frac{B}{x+1} + \frac{C}{x+11}$$

giving

$$4x^2 + 25x + 11 = A(x+1)(x+11) + Bx(x+11) + Cx(x+1)$$
$$4x^2 + 25x + 11 = (A+B+C)x^2 + (12A + 11B + C)x + 11A$$

so

$$A + B + C = 4$$
$$12A + 11B + C = 25$$
$$11A = 11.$$

Thus, $A = B = 1, C = 2$ so

$$\int \frac{x^4 + 12x^3 + 15x^2 + 25x + 11}{x^3 + 12x^2 + 11x}\, dx = \int x\, dx + \int \frac{dx}{x} + \int \frac{dx}{x+1} + \int \frac{2dx}{x+11}$$
$$= \frac{x^2}{2} + \ln|x| + \ln|x+1| + 2\ln|x+11| + K.$$

17. Since $x = \sin t + 2$, we have

$$4x - 3 - x^2 = 4(\sin t + 2) - 3 - (\sin t + 2)^2 = 1 - \sin^2 t = \cos^2 t$$

and $dx = \cos t\, dt$, so substitution gives

$$\int \frac{1}{\sqrt{4x - 3 - x^2}} = \int \frac{1}{\sqrt{\cos^2 t}} \cos t\, dt = \int dt = t + C = \arcsin(x - 2) + C.$$

Problems

21. Since $x^2 + 2x + 2 = (x+1)^2 + 1$, we have

$$\int \frac{1}{x^2 + 2x + 2}\, dx = \int \frac{1}{(x+1)^2 + 1}\, dx.$$

Substitute $x + 1 = \tan\theta$, so $x = (\tan\theta) - 1$.

25. Since $t^2 + 4t + 7 = (t+2)^2 + 3$, we have

$$\int (t+2)\sin(t^2 + 4t + 7)\, dt = \int (t+2)\sin((t+2)^2 + 3)\, dt.$$

Substitute $w = (t+2)^2 + 3$, so $dw = 2(t+2)\, dt$.

This integral can also be computed without completing the square, by substituting $w = t^2 + 4t + 7$, so $dw = (2t + 4)\, dt$.

29. Since $x^3 + x = x(x^2 + 1)$ cannot be factored further, we write

$$\frac{x+1}{x^3 + x} = \frac{A}{x} + \frac{Bx + C}{x^2 + 1}.$$

Multiplying by $x(x^2 + 1)$ gives

$$x + 1 = A(x^2 + 1) + (Bx + C)x$$
$$x + 1 = (A + B)x^2 + Cx + A,$$

so

$$A + B = 0$$
$$C = 1$$
$$A = 1.$$

Thus, $A = C = 1$, $B = -1$, and we have

$$\int \frac{x+1}{x^3+x}\,dx = \int \left(\frac{1}{x} + \frac{-x+1}{x^2+1}\right) = \int \frac{dx}{x} - \int \frac{x\,dx}{x^2+1} + \int \frac{dx}{x^2+1}$$
$$= \ln|x| - \frac{1}{2}\ln\left|x^2+1\right| + \arctan x + K.$$

33. Let $t = \tan\theta$ so $dt = (1/\cos^2\theta)d\theta$. Since $\sqrt{1+\tan^2\theta} = 1/\cos\theta$, we have

$$\int \frac{dt}{t^2\sqrt{1+t^2}} = \int \frac{1/\cos^2\theta}{\tan^2\theta\sqrt{1+\tan^2\theta}}\,d\theta = \int \frac{\cos\theta}{\tan^2\theta\cos^2\theta}\,d\theta = \int \frac{\cos\theta}{\sin^2\theta}\,d\theta.$$

The last integral can be evaluated by guess-and-check or by substituting $w = \sin\theta$. The result is

$$\int \frac{dt}{t^2\sqrt{1+t^2}} = \int \frac{\cos\theta}{\sin^2\theta}\,d\theta = -\frac{1}{\sin\theta} + C.$$

Since $t = \tan\theta$ and $1/\cos^2\theta = 1 + \tan^2\theta$, we have

$$\cos\theta = \frac{1}{\sqrt{1+\tan^2\theta}} = \frac{1}{\sqrt{1+t^2}}.$$

In addition, $\tan\theta = \sin\theta/\cos\theta$ so

$$\sin\theta = \tan\theta\cos\theta = \frac{t}{\sqrt{1+t^2}}.$$

Thus

$$\int \frac{dt}{t^2\sqrt{1+t^2}} = -\frac{\sqrt{1+t^2}}{t} + C.$$

37. Notice that because $\frac{3x}{(x-1)(x-4)}$ is negative for $2 \le x \le 3$,

$$\text{Area} = -\int_2^3 \frac{3x}{(x-1)(x-4)}\,dx.$$

Using partial fractions gives

$$\frac{3x}{(x-1)(x-4)} = \frac{A}{x-1} + \frac{B}{x-4} = \frac{(A+B)x - B - 4A}{(x-1)(x-4)}.$$

Multiplying through by $(x-1)(x-4)$ gives

$$3x = (A+B)x - B - 4A$$

so $A = -1$ and $B = 4$. Thus

$$-\int_2^3 \frac{3x}{(x-1)(x-4)}\,dx = -\int_2^3 \left(\frac{-1}{x-1} + \frac{4}{x-4}\right)dx = \left.(\ln|x-1| - 4\ln|x-4|)\right|_2^3 = 5\ln 2.$$

41. We have

$$\text{Area} = \int_0^3 \frac{1}{\sqrt{x^2+9}}\,dx.$$

Let $x = 3\tan\theta$ so $dx = (3/\cos^2\theta)\,d\theta$ and

$$\sqrt{x^2+9} = \sqrt{\frac{9\sin^2\theta}{\cos^2\theta}+9} = \frac{3}{\cos\theta}.$$

When $x = 0, \theta = 0$ and when $x = 3, \theta = \pi/4$. Thus

$$\int_0^3 \frac{1}{\sqrt{x^2+9}}\,dx = \int_0^{\pi/4} \frac{1}{\sqrt{9\tan^2\theta+9}}\frac{3}{\cos^2\theta}\,d\theta = \int_0^{\pi/4} \frac{1}{3/\cos\theta}\cdot\frac{3}{\cos^2\theta}\,d\theta = \int_0^{\pi/4} \frac{1}{\cos\theta}\,d\theta$$

$$= \frac{1}{2}\ln\left|\frac{\sin\theta+1}{\sin\theta-1}\right|\bigg|_0^{\pi/4} = \frac{1}{2}\ln\left|\frac{1/\sqrt{2}+1}{1/\sqrt{2}-1}\right| = \frac{1}{2}\ln\left(\frac{1+\sqrt{2}}{\sqrt{2}-1}\right).$$

This answer can be simplified to $\ln(1+\sqrt{2})$ by multiplying the numerator and denominator of the fraction by $(\sqrt{2}+1)$ and using the properties of logarithms. The integral $\int (1/\cos\theta)\,d\theta$ is done using the Table of Integrals.

45. (a) We want to evaluate the integral

$$T = \int_0^{a/2} \frac{k\,dx}{(a-x)(b-x)}.$$

Using partial fractions, we have

$$\frac{k}{(a-x)(b-x)} = \frac{C}{a-x} + \frac{D}{b-x}$$
$$k = C(b-x) + D(a-x)$$
$$k = -(C+D)x + Cb + Da$$

so

$$0 = -(C+D)$$
$$k = Cb + Da,$$

giving

$$C = -D = \frac{k}{b-a}.$$

Thus, the time is given by

$$T = \int_0^{a/2} \frac{k\,dx}{(a-x)(b-x)} = \frac{k}{b-a}\int_0^{a/2}\left(\frac{1}{a-x} - \frac{1}{b-x}\right)dx$$

$$= \frac{k}{b-a}\left(-\ln|a-x| + \ln|b-x|\right)\bigg|_0^{a/2}$$

$$= \frac{k}{b-a}\ln\left|\frac{b-x}{a-x}\right|\bigg|_0^{a/2}$$

$$= \frac{k}{b-a}\left(\ln\left(\frac{2b-a}{a}\right) - \ln\left(\frac{b}{a}\right)\right)$$

$$= \frac{k}{b-a}\ln\left(\frac{2b-a}{b}\right).$$

(b) A similar calculation with x_0 instead of $a/2$ leads to the following expression for the time

$$T = \int_0^{x_0} \frac{k\,dx}{(a-x)(b-x)} = \frac{k}{b-a}\ln\left|\frac{b-x}{a-x}\right|\bigg|_0^{x_0}$$

$$= \frac{k}{b-a}\left(\ln\left|\frac{b-x_0}{a-x_0}\right| - \ln\left(\frac{b}{a}\right)\right).$$

As $x_0 \to a$, the value of $|a-x_0| \to 0$, so $|b-x_0|/|a-x_0| \to \infty$. Thus, $T \to \infty$ as $x_0 \to a$. In other words, the time taken tends to infinity.

Solutions for Section 7.5

Exercises

1. **(a)** The approximation LEFT(2) uses two rectangles, with the height of each rectangle determined by the left-hand endpoint. See Figure 7.2. We see that this approximation is an underestimate.

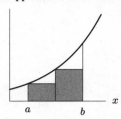

Figure 7.2

(b) The approximation RIGHT(2) uses two rectangles, with the height of each rectangle determined by the right-hand endpoint. See Figure 7.3. We see that this approximation is an overestimate.

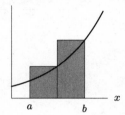

Figure 7.3

(c) The approximation TRAP(2) uses two trapezoids, with the height of each trapezoid given by the secant line connecting the two endpoints. See Figure 7.4. We see that this approximation is an overestimate.

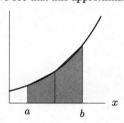

Figure 7.4

(d) The approximation MID(2) uses two rectangles, with the height of each rectangle determined by the height at the midpoint. Alternately, we can view MID(2) as a trapezoid rule where the height is given by the tangent line at the midpoint. Both interpretations are shown in Figure 7.5. We see from the tangent line interpretation that this approximation is an underestimate

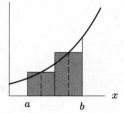
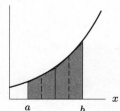

Figure 7.5

5. **(a)** Since two rectangles are being used, the width of each rectangle is 3. The height is given by the left-hand endpoint so we have

$$\text{LEFT}(2) = f(0) \cdot 3 + f(3) \cdot 3 = 0^2 \cdot 3 + 3^2 \cdot 3 = 27.$$

(b) Since two rectangles are being used, the width of each rectangle is 3. The height is given by the right-hand endpoint so we have

$$\text{RIGHT}(2) = f(3) \cdot 3 + f(6) \cdot 3 = 3^2 \cdot 3 + 6^2 \cdot 3 = 135.$$

(c) We know that TRAP is the average of LEFT and RIGHT and so

$$\text{TRAP}(2) = \frac{27 + 135}{2} = 81.$$

(d) Since two rectangles are being used, the width of each rectangle is 3. The height is given by the height at the midpoint so we have

$$\text{MID}(2) = f(1.5) \cdot 3 + f(4.5) \cdot 3 = (1.5)^2 \cdot 3 + (4.5)^2 \cdot 3 = 67.5.$$

Problems

9. Let $s(t)$ be the distance traveled at time t and $v(t)$ be the velocity at time t. Then the distance traveled during the interval $0 \le t \le 6$ is

$$s(6) - s(0) = s(t)\Big|_0^6$$
$$= \int_0^6 s'(t)\, dt \quad \text{(by the Fundamental Theorem)}$$
$$= \int_0^6 v(t)\, dt.$$

We estimate the distance by estimating this integral.

From the table, we find: $\text{LEFT}(6) = 31$, $\text{RIGHT}(6) = 39$, $\text{TRAP}(6) = 35$.

13. $f(x)$ is concave down, so MID gives an overestimate and TRAP gives an underestimate.

17. (a) TRAP(4) gives probably the best estimate of the integral. We cannot calculate MID(4).

$$\text{LEFT}(4) = 3 \cdot 100 + 3 \cdot 97 + 3 \cdot 90 + 3 \cdot 78 = 1095$$
$$\text{RIGHT}(4) = 3 \cdot 97 + 3 \cdot 90 + 3 \cdot 78 + 3 \cdot 55 = 960$$
$$\text{TRAP}(4) = \frac{1095 + 960}{2} = 1027.5.$$

(b) Because there are no points of inflection, the graph is either concave down or concave up. By plotting points, we see that it is concave down. So TRAP(4) is an underestimate.

21.

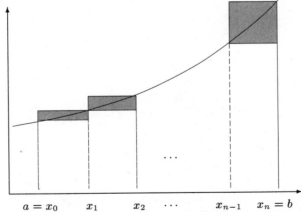

From the diagram, the difference between $\text{RIGHT}(n)$ and $\text{LEFT}(n)$ is the area of the shaded rectangles.

$$\text{RIGHT}(n) = f(x_1)\Delta x + f(x_2)\Delta x + \cdots + f(x_n)\Delta x$$
$$\text{LEFT}(n) = f(x_0)\Delta x + f(x_1)\Delta x + \cdots + f(x_{n-1})\Delta x$$

Notice that the terms in these two sums are the same, except that $\text{RIGHT}(n)$ contains $f(x_n)\Delta x\ (= f(b)\Delta x)$, and $\text{LEFT}(n)$ contains $f(x_0)\Delta x\ (= f(a)\Delta x)$. Thus

$$\text{RIGHT}(n) = \text{LEFT}(n) + f(x_n)\Delta x - f(x_0)\Delta x$$
$$= \text{LEFT}(n) + f(b)\Delta x - f(a)\Delta x$$

25. First, we compute:

$$(f(b) - f(a))\Delta x = (f(b) - f(a))\left(\frac{b-a}{n}\right)$$

$$= (f(5) - f(2))\left(\frac{3}{n}\right)$$

$$= (21 - 13)\left(\frac{3}{n}\right)$$

$$= \frac{24}{n}$$

RIGHT(10) = LEFT(10) + 24 = 3.156 + 2.4 = 5.556.
TRAP(10) = LEFT(10) + $\frac{1}{2}$(2.4) = 3.156 + 1.2 = 4.356.
LEFT(20) = $\frac{1}{2}$(LEFT(10) + MID(10)) = $\frac{1}{2}$(3.156 + 3.242) = 3.199.
RIGHT(20) = LEFT(20) + 2.4 = 3.199 + 1.2 = 4.399.
TRAP(20) = LEFT(20) + $\frac{1}{2}$(1.2) = 3.199 + 0.6 = 3.799.

Solutions for Section 7.6

Exercises

1. We saw in Problem 5 in Section 7.5 that, for this definite integral, we have LEFT(2) = 27, RIGHT(2) = 135, TRAP(2) = 81, and MID(2) = 67.5. Thus,

$$\text{SIMP}(2) = \frac{2\text{MID}(2) + \text{TRAP}(2)}{3} = \frac{2(67.5) + 81}{3} = 72.$$

Notice that

$$\int_0^6 x^2 \, dx = \left.\frac{x^3}{3}\right|_0^6 = \frac{6^3}{3} - \frac{0^3}{3} = 72,$$

and so SIMP(2) gives the exact value of the integral in this case.

Problems

5. (a) $\int_0^4 e^x \, dx = \left. e^x \right|_0^4 = e^4 - e^0 \approx 53.598\ldots.$

(b) Computing the sums directly, since $\Delta x = 2$, we have
LEFT(2)= $2 \cdot e^0 + 2 \cdot e^2 \approx 2(1) + 2(7.389) = 16.778;$ error = 36.820.
RIGHT(2)= $2 \cdot e^2 + 2 \cdot e^4 \approx 2(7.389) + 2(54.598) = 123.974;$ error = −70.376.
TRAP(2)= $\dfrac{16.778 + 123.974}{2} = 70.376;$ error = 16.778.
MID(2)= $2 \cdot e^1 + 2 \cdot e^3 \approx 2(2.718) + 2(20.086) = 45.608;$ error = 7.990.
SIMP(2)= $\dfrac{2(45.608) + 70.376}{3} = 53.864;$ error = −0.266.

(c) Similarly, since $\Delta x = 1$, we have LEFT(4)= 31.193;   error = 22.405
RIGHT(4)= 84.791;   error = −31.193
TRAP(4)= 57.992;   error = −4.394
MID(4)= 51.428;   error = 2.170
SIMP(4)= 53.616;   error = −0.018

(d) For LEFT and RIGHT, we expect the error to go down by 1/2, and this is very roughly what we see. For MID and TRAP, we expect the error to go down by 1/4, and this is approximately what we see. For SIMP, we expect the error to go down by $1/2^4 = 1/16$, and this is approximately what we see.

9. Since the midpoint rule is sensitive to f'', the simplifying assumption should be that f'' does not change sign in the interval of integration. Thus MID(10) and MID(20) will both be overestimates or will both be underestimates. Since the larger number, MID(10) is less accurate than the smaller number, they must both be overestimates. Then the information that ERROR(10) $= 4 \times$ ERROR(20) means that the the value of the integral and the two sums are arranged as follows:

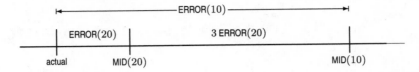

Thus

$$3 \times \text{ERROR}(20) = \text{MID}(10) - \text{MID}(20) = 35.619 - 35.415 = 0.204,$$

so ERROR(20) $= 0.068$ and ERROR(10) $= 4 \times$ ERROR(20) $= 0.272$.

Solutions for Section 7.7

Exercises

1. (a) See Figure 7.6. The area extends out infinitely far along the positive x-axis.

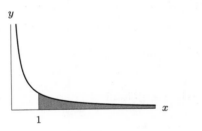

Figure 7.6

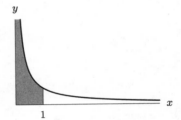

Figure 7.7

(b) See Figure 7.7. The area extends up infinitely far along the positive y-axis.

5. We have

$$\int_1^\infty \frac{1}{5x+2}\,dx = \lim_{b\to\infty} \int_1^b \frac{1}{5x+2}\,dx = \lim_{b\to\infty} \left(\frac{1}{5}\ln(5x+2)\right)\Big|_1^b = \lim_{b\to\infty} \left(\frac{1}{5}\ln(5b+2) - \frac{1}{5}\ln(7)\right).$$

As $b \leftarrow \infty$, we know that $\ln(5b+2) \to \infty$, and so this integral diverges.

9. Using integration by parts with $u = x$ and $v' = e^{-x}$, we find that

$$\int xe^{-x}\,dx = -xe^{-x} - \int -e^{-x}\,dx = -(1+x)e^{-x}$$

so

$$\int_0^\infty \frac{x}{e^x}\,dx = \lim_{b\to\infty} \int_0^b \frac{x}{e^x}\,dx$$

$$= \lim_{b\to\infty} -1(1+x)e^{-x}\Big|_0^b$$

$$= \lim_{b\to\infty} \left[1 - (1+b)e^{-b}\right]$$

$$= 1.$$

13. This is an improper integral because $\sqrt{16 - x^2} = 0$ at $x = 4$. So

$$\int_0^4 \frac{dx}{\sqrt{16 - x^2}} = \lim_{b \to 4^-} \int_0^b \frac{dx}{\sqrt{16 - x^2}}$$

$$= \lim_{b \to 4^-} (\arcsin x/4) \Big|_0^b$$

$$= \lim_{b \to 4^-} [\arcsin(b/4) - \arcsin(0)] = \pi/2 - 0 = \pi/2.$$

17.

$$\int_1^\infty \frac{1}{x^2 + 1} \, dx = \lim_{b \to \infty} \int_1^b \frac{1}{x^2 + 1} \, dx$$

$$= \lim_{b \to \infty} \arctan(x) \Big|_1^b$$

$$= \lim_{b \to \infty} [\arctan(b) - \arctan(1)]$$

$$= \pi/2 - \pi/4 = \pi/4.$$

21. With the substitution $w = \ln x$, $dw = \frac{1}{x} dx$,

$$\int \frac{dx}{x \ln x} = \int \frac{1}{w} \, dw = \ln |w| + C = \ln |\ln x| + C$$

so

$$\int_2^\infty \frac{dx}{x \ln x} = \lim_{b \to \infty} \int_2^b \frac{dx}{x \ln x}$$

$$= \lim_{b \to \infty} \ln |\ln x| \Big|_2^b$$

$$= \lim_{b \to \infty} [\ln |\ln b| - \ln |\ln 2|].$$

As $b \to \infty$, the limit goes to ∞ and hence the integral diverges.

25. Using the substitution $w = -x^{\frac{1}{2}}$, $-2dw = x^{-\frac{1}{2}} dx$,

$$\int e^{-x^{\frac{1}{2}}} x^{-\frac{1}{2}} \, dx = -2 \int e^w \, dw = -2e^{-x^{\frac{1}{2}}} + C.$$

So

$$\int_0^\pi \frac{1}{\sqrt{x}} e^{-\sqrt{x}} \, dx = \lim_{b \to 0^+} \int_b^\pi \frac{1}{\sqrt{x}} e^{-\sqrt{x}} \, dx$$

$$= \lim_{b \to 0^+} -2e^{-\sqrt{x}} \Big|_b^\pi$$

$$= 2 - 2e^{-\sqrt{\pi}}.$$

29. $\int \frac{dx}{x^2 - 1} = \int \frac{dx}{(x-1)(x+1)} = \frac{1}{2}(\ln |x - 1| - \ln |x + 1|) + C = \frac{1}{2} \left(\ln \frac{|x - 1|}{|x + 1|} \right) + C$, so

$$\int_4^\infty \frac{dx}{x^2 - 1} = \lim_{b \to \infty} \int_4^b \frac{dx}{x^2 - 1}$$

$$= \lim_{b \to \infty} \frac{1}{2} \left(\ln \frac{|x - 1|}{|x + 1|} \right) \Big|_4^b$$

$$= \lim_{b \to \infty} \left[\frac{1}{2} \ln \left(\frac{b - 1}{b + 1} \right) - \frac{1}{2} \ln \frac{3}{5} \right]$$

$$= -\frac{1}{2} \ln \frac{3}{5} = \frac{1}{2} \ln \frac{5}{3}.$$

Problems

33. Since the graph is above the x-axis for $x \geq 0$, we have

$$
\text{Area} = \int_0^\infty xe^{-x}\, dx = \lim_{b \to \infty} \int_0^b xe^{-x}\, dx
$$

$$
= \lim_{b \to \infty} \left(-xe^{-x} \Big|_0^b + \int_0^b e^{-x}\, dx \right)
$$

$$
= \lim_{b \to \infty} \left(-be^{-b} - e^{-x} \Big|_0^b \right)
$$

$$
= \lim_{b \to \infty} (-be^{-b} - e^{-b} + e^0) = 1.
$$

37. The factor $\ln x$ grows slowly enough (as $x \to 0^+$) not to change the convergence or divergence of the integral, although it will change what it converges or diverges to.

The integral is always improper, because $\ln x$ is not defined for $x = 0$. Integrating by parts (or, alternatively, the integral table) yields

$$
\int_0^e x^p \ln x\, dx = \lim_{a \to 0^+} \int_a^e x^p \ln x\, dx
$$

$$
= \lim_{a \to 0^+} \left(\frac{1}{p+1} x^{p+1} \ln x - \frac{1}{(p+1)^2} x^{p+1} \right) \Big|_a^e
$$

$$
= \lim_{a \to 0^+} \left[\left(\frac{1}{p+1} e^{p+1} - \frac{1}{(p+1)^2} e^{p+1} \right) \right.
$$

$$
\left. - \left(\frac{1}{p+1} a^{p+1} \ln a - \frac{1}{(p+1)^2} a^{p+1} \right) \right].
$$

If $p < -1$, then $(p+1)$ is negative, so as $a \to 0^+$, $a^{p+1} \to \infty$ and $\ln a \to -\infty$, and therefore the limit does not exist.

If $p > -1$, then $(p+1)$ is positive and it's easy to see that $a^{p+1} \to 0$ as $a \to 0$. Looking at graphs of $x^{p+1} \ln x$ (for different values of p) shows that $a^{p+1} \ln a \to 0$ as $a \to 0$. This isn't so easy to see analytically. It's true because if we let $t = \frac{1}{a}$ then

$$
\lim_{a \to 0^+} a^{p+1} \ln a = \lim_{t \to \infty} \left(\frac{1}{t} \right)^{p+1} \ln \left(\frac{1}{t} \right) = \lim_{t \to \infty} -\frac{\ln t}{t^{p+1}}.
$$

This last limit is zero because $\ln t$ grows very slowly, much more slowly than t^{p+1}. So if $p > -1$, the integral converges and equals $e^{p+1}[1/(p+1) - 1/(p+1)^2] = pe^{p+1}/(p+1)^2$.

What happens if $p = -1$? Then we get

$$
\int_0^e \frac{\ln x}{x}\, dx = \lim_{a \to 0^+} \int_a^e \frac{\ln x}{x}\, dx
$$

$$
= \lim_{a \to 0^+} \frac{(\ln x)^2}{2} \Big|_a^e
$$

$$
= \lim_{a \to 0^+} \left(\frac{1 - (\ln a)^2}{2} \right).
$$

Since $\ln a \to -\infty$ as $a \to 0^+$, this limit does not exist.

To summarize, $\int_0^e x^p \ln x$ converges for $p > -1$ to the value $pe^{p+1}/(p+1)^2$.

Solutions for Section 7.8

Exercises

1. For large x, the integrand behaves like $1/x^2$ because

$$\frac{x^2}{x^4 + 1} \approx \frac{x^2}{x^4} = \frac{1}{x^2}.$$

Since $\int_1^\infty \frac{dx}{x^2}$ converges, we expect our integral to converge. More precisely, since $x^4 + 1 > x^4$, we have

$$\frac{x^2}{x^4 + 1} < \frac{x^2}{x^4} = \frac{1}{x^2}.$$

Since $\int_1^\infty \frac{dx}{x^2}$ is convergent, the comparison test tells us that $\int_1^\infty \frac{x^2}{x^4 + 1} \, dx$ converges also.

5. The integrand is continuous for all $x \geq 1$, so whether the integral converges or diverges depends only on the behavior of the function as $x \to \infty$. As $x \to \infty$, polynomials behave like the highest powered term. Thus, as $x \to \infty$, the integrand $\frac{x}{x^2 + 2x + 4}$ behaves like $\frac{x}{x^2}$ or $\frac{1}{x}$. Since $\int_1^\infty \frac{1}{x} \, dx$ diverges, we predict that the given integral will diverge.

9. The integrand is continuous for all $x \geq 1$, so whether the integral converges or diverges depends only on the behavior of the function as $x \to \infty$. As $x \to \infty$, polynomials behave like the highest powered term. Thus, as $x \to \infty$, the integrand $\frac{x^2 + 4}{x^4 + 3x^2 + 11}$ behaves like $\frac{x^2}{x^4}$ or $\frac{1}{x^2}$. Since $\int_1^\infty \frac{1}{x^2} \, dx$ converges, we predict that the given integral will converge.

13. The integrand is unbounded as $t \to 5$. We substitute $w = t - 5$, so $dw = dt$. When $t = 5$, $w = 0$ and when $t = 8$, $w = 3$.

$$\int_5^8 \frac{6}{\sqrt{t - 5}} \, dt = \int_0^3 \frac{6}{\sqrt{w}} \, dw.$$

Since

$$\int_0^3 \frac{6}{\sqrt{w}} \, dw = \lim_{a \to 0^+} 6 \int_a^3 \frac{1}{\sqrt{w}} \, dw = 6 \lim_{a \to 0^+} 2w^{1/2} \Big|_a^3 = 12 \lim_{a \to 0^+} (\sqrt{3} - \sqrt{a}) = 12\sqrt{3},$$

our integral converges.

17. Since $\frac{1}{u + u^2} < \frac{1}{u^2}$ for $u \geq 1$, and since $\int_1^\infty \frac{du}{u^2}$ converges, $\int_1^\infty \frac{du}{u + u^2}$ converges.

21. Since $\frac{1}{1 + e^y} \leq \frac{1}{e^y} = e^{-y}$ and $\int_0^\infty e^{-y} \, dy$ converges, the integral $\int_0^\infty \frac{dy}{1 + e^y}$ converges.

25. Since $\frac{3 + \sin \alpha}{\alpha} \geq \frac{2}{\alpha}$ for $\alpha \geq 4$, and since $\int_4^\infty \frac{2}{\alpha} d\alpha$ diverges, then $\int_4^\infty \frac{3 + \sin \alpha}{\alpha} d\alpha$ diverges.

Problems

29. First let's calculate the indefinite integral $\int \frac{dx}{x(\ln x)^p}$. Let $\ln x = w$, then $\frac{dx}{x} = dw$. So

$$\int \frac{dx}{x(\ln x)^p} = \int \frac{dw}{w^p}$$

$$= \begin{cases} \ln |w| + C, & \text{if } p = 1 \\ \frac{1}{1-p} w^{1-p} + C, & \text{if } p \neq 1 \end{cases}$$

$$= \begin{cases} \ln |\ln x| + C, & \text{if } p = 1 \\ \frac{1}{1-p} (\ln x)^{1-p} + C, & \text{if } p \neq 1. \end{cases}$$

Notice that $\lim_{x \to \infty} \ln x = +\infty$.

(a) $p = 1$:

$$\int_2^\infty \frac{dx}{x \ln x} = \lim_{b \to \infty} \left(\ln |\ln b| - \ln |\ln 2| \right) = +\infty.$$

(b) $p < 1$:

$$\int_2^\infty \frac{dx}{x(\ln x)^p} = \frac{1}{1-p} \left(\lim_{b \to \infty} (\ln b)^{1-p} - (\ln 2)^{1-p} \right) = +\infty.$$

(c) $p > 1$:

$$\int_2^\infty \frac{dx}{x(\ln x)^p} = \frac{1}{1-p} \left(\lim_{b \to \infty} (\ln b)^{1-p} - (\ln 2)^{1-p} \right)$$

$$= \frac{1}{1-p} \left(\lim_{b \to \infty} \frac{1}{(\ln b)^{p-1}} - (\ln 2)^{1-p} \right)$$

$$= -\frac{1}{1-p} (\ln 2)^{1-p}.$$

Thus, $\displaystyle\int_2^\infty \frac{dx}{x(\ln x)^p}$ is convergent for $p > 1$, divergent for $p \leq 1$.

33. (a) Since $e^{-x^2} \leq e^{-3x}$ for $x \geq 3$,

$$\int_3^\infty e^{-x^2} \, dx \leq \int_3^\infty e^{-3x} \, dx$$

Now

$$\int_3^\infty e^{-3x} \, dx = \lim_{b \to \infty} \int_3^b e^{-3x} \, dx = \lim_{b \to \infty} -\frac{1}{3} e^{-3x} \bigg|_3^b$$

$$= \lim_{b \to \infty} \frac{e^{-9}}{3} - \frac{e^{-3b}}{3} = \frac{e^{-9}}{3}.$$

Thus

$$\int_3^\infty e^{-x^2} \, dx \leq \frac{e^{-9}}{3}.$$

(b) By reasoning similar to part (a),

$$\int_n^\infty e^{-x^2} \, dx \leq \int_n^\infty e^{-nx} \, dx,$$

and

$$\int_n^\infty e^{-nx} \, dx = \frac{1}{n} e^{-n^2},$$

so

$$\int_n^\infty e^{-x^2} \, dx \leq \frac{1}{n} e^{-n^2}.$$

Solutions for Chapter 7 Review

Exercises

1. Since $\dfrac{d}{dt} \cos t = -\sin t$, we have

$$\int \sin t \, dt = -\cos t + C, \quad \text{where } C \text{ is a constant.}$$

5. Since $\displaystyle\int \sin w \, d\theta = -\cos w + C$, the substitution $w = 2\theta$, $dw = 2 \, d\theta$ gives $\displaystyle\int \sin 2\theta \, d\theta = -\frac{1}{2} \cos 2\theta + C.$

9. Either expand $(r+1)^3$ or use the substitution $w = r+1$. If $w = r+1$, then $dw = dr$ and

$$\int (r+1)^3 \, dr = \int w^3 \, dw = \frac{1}{4} w^4 + C = \frac{1}{4}(r+1)^4 + C.$$

13. Substitute $w = t^2$, so $dw = 2t \, dt$.

$$\int te^{t^2} \, dt = \frac{1}{2} \int e^{t^2} 2t \, dt = \frac{1}{2} \int e^w \, dw = \frac{1}{2} e^w + C = \frac{1}{2} e^{t^2} + C.$$

Check:

$$\frac{d}{dt}\left(\frac{1}{2}e^{t^2} + C\right) = 2t\left(\frac{1}{2}e^{t^2}\right) = te^{t^2}.$$

17. Integration by parts with $u = \ln x$, $v' = x$ gives

$$\int x \ln x \, dx = \frac{x^2}{2} \ln x - \int \frac{1}{2} x \, dx = \frac{1}{2} x^2 \ln x - \frac{1}{4} x^2 + C.$$

Or use the integral table, III-13, with $n = 1$.

21. Using the exponent rules and the chain rule, we have

$$\int e^{0.5 - 0.3t} \, dt = e^{0.5} \int e^{-0.3t} \, dt = -\frac{e^{0.5}}{0.3} e^{-0.3t} + C = -\frac{e^{0.5 - 0.3t}}{0.3} + C.$$

25. Substitute $w = \sqrt{y}$, $dw = 1/(2\sqrt{y}) \, dy$. Then

$$\int \frac{\cos \sqrt{y}}{\sqrt{y}} \, dy = 2 \int \cos w \, dw = 2 \sin w + C = 2 \sin \sqrt{y} + C.$$

Check:

$$\frac{d}{dy} 2 \sin \sqrt{y} + C = \frac{2 \cos \sqrt{y}}{2\sqrt{y}} = \frac{\cos \sqrt{y}}{\sqrt{y}}.$$

29. Substitute $w = 2x - 6$. Then $dw = 2 \, dx$ and

$$\int \tan(2x - 6) \, dx = \frac{1}{2} \int \tan w \, dw = \frac{1}{2} \int \frac{\sin w}{\cos w} \, dw$$

$$= -\frac{1}{2} \ln |\cos w| + C \text{ by substitution or by I-7 of the integral table.}$$

$$= -\frac{1}{2} \ln |\cos(2x - 6)| + C.$$

33.

$$\int_0^{10} ze^{-z} \, dz = [-ze^{-z}]\Big|_0^{10} - \int_0^{10} -e^{-z} \, dz \qquad (\text{let } z = u, e^{-z} = v', -e^{-z} = v)$$

$$= -10e^{-10} - [e^{-z}]\Big|_0^{10}$$

$$= -10e^{-10} - e^{-10} + 1$$

$$= -11e^{-10} + 1.$$

37.

$$\int_0^1 \frac{dx}{x^2 + 1} = \tan^{-1} x \Big|_0^1 = \tan^{-1} 1 - \tan^{-1} 0 = \frac{\pi}{4} - 0 = \frac{\pi}{4}.$$

41. Dividing and then integrating, we obtain

$$\int \frac{t + 1}{t^2} \, dt = \int \frac{1}{t} \, dt + \int \frac{1}{t^2} \, dt = \ln |t| - \frac{1}{t} + C, \text{ where } C \text{ is a constant.}$$

45. Using substitution,

$$\int \frac{x}{x^2+1}\,dx = \int \frac{1/2}{w}\,dw \qquad (x^2+1=w, 2x\,dx=dw, x\,dx=\frac{1}{2}\,dw)$$

$$= \frac{1}{2}\int \frac{1}{w}\,dw = \frac{1}{2}\ln|w| + C = \frac{1}{2}\ln|x^2+1| + C,$$

where C is a constant.

49. Let $\cos 5\theta = w$, then $-5\sin 5\theta\,d\theta = dw$, $\sin 5\theta\,d\theta = -\frac{1}{5}dw$. So

$$\int \sin 5\theta \cos^3 5\theta\,d\theta = \int w^3 \cdot \left(-\frac{1}{5}\right)dw = -\frac{1}{5}\int w^3\,dw = -\frac{1}{20}w^4 + C$$

$$= -\frac{1}{20}\cos^4 5\theta + C,$$

where C is a constant.

53.

$$\int xe^x\,dx = xe^x - \int e^x\,dx \qquad (\text{let } x=u, e^x=v', e^x=v)$$

$$= xe^x - e^x + C,$$

where C is a constant.

57. Rewrite $9+u^2$ as $9[1+(u/3)^2]$ and let $w=u/3$, then $dw=du/3$ so that

$$\int \frac{du}{9+u^2} = \frac{1}{3}\int \frac{dw}{1+w^2} = \frac{1}{3}\arctan w + C = \frac{1}{3}\arctan\left(\frac{u}{3}\right) + C.$$

61. Let $u=2x$, then $du=2\,dx$ so that

$$\int \frac{dx}{\sqrt{1-4x^2}} = \frac{1}{2}\int \frac{du}{\sqrt{1-u^2}} = \frac{1}{2}\arcsin u + C = \frac{1}{2}\arcsin(2x) + C.$$

65. Let $w=\ln x$. Then $dw=(1/x)dx$ which gives

$$\int \frac{dx}{x\ln x} = \int \frac{dw}{w} = \ln|w| + C = \ln|\ln x| + C.$$

69. Using integration by parts, let $r=u$ and $dt=e^{ku}du$, so $dr=du$ and $t=(1/k)e^{ku}$. Thus

$$\int ue^{ku}\,du = \frac{u}{k}e^{ku} - \frac{1}{k}\int e^{ku}\,du = \frac{u}{k}e^{ku} - \frac{1}{k^2}e^{ku} + C.$$

73. $\int (e^x+x)^2\,dx = \int (e^{2x}+2xe^x+x^2)\,dy$. Separating into three integrals, we have

$$\int e^{2x}\,dx = \frac{1}{2}\int e^{2x}2\,dx = \frac{1}{2}e^{2x} + C_1,$$

$$\int 2xe^x\,dx = 2\int xe^x\,dx = 2xe^x - 2e^x + C_2$$

from Formula II-13 of the integral table or integration by parts, and

$$\int x^2\,dx = \frac{x^3}{3} + C_3.$$

Combining the results and writing $C=C_1+C_2+C_3$, we get

$$\frac{1}{2}e^{2x} + 2xe^x - 2e^x + \frac{x^3}{3} + C.$$

77. We can factor $r^2-100 = (r-10)(r+10)$ so we can use Table V-26 (with $a=10$ and $b=-10$) to get

$$\int \frac{dr}{r^2-100} = \frac{1}{20}\left[\ln|r-10| + \ln|r+10|\right] + C.$$

81. Since $\int (x^{\sqrt{k}} + (\sqrt{k})^x)dx = \int x^{\sqrt{k}} dx + \int (\sqrt{k})^x \, dx$, for the first integral, use Formula I-1 with $n = \sqrt{k}$. For the second integral, use Formula I-3 with $a = \sqrt{k}$. The result is

$$\int (x^{\sqrt{x}} + (\sqrt{k})^x) \, dx = \frac{x^{(\sqrt{k})+1}}{(\sqrt{k}) + 1} + \frac{(\sqrt{k})^x}{\ln \sqrt{k}} + C.$$

85. First divide $x^2 + 3x + 2$ into x^3 to obtain

$$\frac{x^3}{x^2 + 3x + 2} = x - 3 + \frac{7x + 6}{x^2 + 3x + 2}.$$

Since $x^2 + 3x + 2 = (x + 1)(x + 2)$, we can use V-27 of the integral table (with $c = 7$, $d = 6$, $a = -1$, and $b = -2$) to get

$$\int \frac{7x + 6}{x^2 + 3x + 2} \, dx = -\ln|x + 1| + 8\ln|x + 2| + C.$$

Including the terms $x - 3$ from the long division and integrating them gives

$$\int \frac{x^3}{x^2 + 3x + 2} \, dx = \int \left(x - 3 + \frac{7x + 6}{x^2 + 3x + 6} \right) \, dx = \frac{1}{2}x^2 - 3x - \ln|x + 1| + 8\ln|x + 2| + C.$$

89. This can be done by formula V–26 in the integral table or by partial fractions

$$\int \frac{dz}{z^2 + z} = \int \frac{dz}{z(z + 1)} = \int \left(\frac{1}{z} - \frac{1}{z + 1} \right) \, dz = \ln|z| - \ln|z + 1| + C.$$

Check:

$$\frac{d}{dz} \left(\ln|z| - \ln|z + 1| + C \right) = \frac{1}{z} - \frac{1}{z + 1} = \frac{1}{z^2 + z}.$$

93. Multiplying out and integrating term by term gives

$$\int (x^2 + 5)^3 dx = \int (x^6 + 15x^4 + 75x^2 + 125)dx = \frac{1}{7}x^7 + 15\frac{x^5}{5} + 75\frac{x^3}{3} + 125x + C$$

$$= \frac{1}{7}x^7 + 3x^5 + 25x^3 + 125x + C.$$

97. If $u = 1 + \cos^2 w$, $du = 2(\cos w)^1(-\sin w) \, dw$, so

$$\int \frac{\sin w \cos w}{1 + \cos^2 w} \, dw = -\frac{1}{2} \int \frac{-2\sin w \cos w}{1 + \cos^2 w} \, dw = -\frac{1}{2} \int \frac{1}{u} \, du = -\frac{1}{2} \ln|u| + C$$

$$= -\frac{1}{2} \ln|1 + \cos^2 w| + C.$$

101. If $u = \sqrt{x + 1}$, $u^2 = x + 1$ with $x = u^2 - 1$ and $dx = 2u \, du$. Substituting, we get

$$\int \frac{x}{\sqrt{x + 1}} \, dx = \int \frac{(u^2 - 1)2u \, du}{u} = \int (u^2 - 1)2 \, du = 2 \int (u^2 - 1) \, du$$

$$= \frac{2u^3}{3} - 2u + C = \frac{2(\sqrt{x + 1})^3}{3} - 2\sqrt{x + 1} + C.$$

105. Letting $u = z - 5$, $z = u + 5$, $dz = du$, and substituting, we have

$$\int \frac{z}{(z - 5)^3} dz = \int \frac{u + 5}{u^3} du = \int (u^{-2} + 5u^{-3})du = \frac{u^{-1}}{-1} + 5\left(\frac{u^{-2}}{-2} \right) + C$$

$$= \frac{-1}{(z - 5)} + \frac{-5}{2(z - 5)^2} + C.$$

109. Let $w = 2 + 3\cos x$, so $dw = -3\sin x\, dx$, giving $-\dfrac{1}{3}\, dw = \sin x\, dx$. Then

$$\int \sin x\left(\sqrt{2 + 3\cos x}\right) dx = \int \sqrt{w}\left(-\frac{1}{3}\right) dw = -\frac{1}{3}\int \sqrt{w}\, dw$$

$$= \left(-\frac{1}{3}\right)\frac{w^{\frac{3}{2}}}{\frac{3}{2}} + C = -\frac{2}{9}(2 + 3\cos x)^{\frac{3}{2}} + C.$$

113. Let $w = x + \sin x$, then $dw = (1 + \cos x)\, dx$ which gives

$$\int (x + \sin x)^3 (1 + \cos x)\, dx = \int w^3\, dw = \frac{1}{4}w^4 + C = \frac{1}{4}(x + \sin x)^4 + C.$$

117. Splitting the integrand into partial fractions with denominators x and $(x + 5)$, we have

$$\frac{1}{x(x + 5)} = \frac{A}{x} + \frac{B}{x + 5}.$$

Multiplying by $x(x + 5)$ gives the identity

$$1 = A(x + 5) + Bx$$

so

$$1 = (A + B)x + 5A.$$

Since this equation holds for all x, the constant terms on both sides must be equal. Similarly, the coefficient of x on both sides must be equal. So

$$5A = 1$$
$$A + B = 0.$$

Solving these equations gives $A = 1/5$, $B = -1/5$ and the integral becomes

$$\int \frac{1}{x(x + 5)}\, dx = \frac{1}{5}\int \frac{1}{x}\, dx - \frac{1}{5}\int \frac{1}{x + 5}\, dx = \frac{1}{5}\left(\ln |x| - \ln |x + 5|\right) + C.$$

121. The denominator can be factored to give $x(x - 1)(x + 1)$. Splitting the integrand into partial fractions with denominators x, $x - 1$, and $x + 1$, we have

$$\frac{3x + 1}{x(x - 1)(x + 1)} = \frac{A}{x - 1} + \frac{B}{x + 1} + \frac{C}{x}.$$

Multiplying by $x(x - 1)(x + 1)$ gives the identity

$$3x + 1 = Ax(x + 1) + Bx(x - 1) + C(x - 1)(x + 1)$$

so

$$3x + 1 = (A + B + C)x^2 + (A - B)x - C.$$

Since this equation holds for all x, the constant terms on both sides must be equal. Similarly, the coefficient of x and x^2 on both sides must be equal. So

$$-C = 1$$
$$A - B = 3$$
$$A + B + C = 0.$$

Solving these equations gives $A = 2$, $B = -1$ and $C = -1$. The integral becomes

$$\int \frac{3x + 1}{x(x + 1)(x - 1)}\, dx = \int \frac{2}{x - 1}\, dx - \int \frac{1}{x + 1}\, dx - \int \frac{1}{x}\, dx$$

$$= 2\ln |x - 1| - \ln |x + 1| - \ln |x| + C.$$

125. Using partial fractions, we have:

$$\frac{3x+1}{x^2-3x+2} = \frac{3x+1}{(x-1)(x-2)} = \frac{A}{x-1} + \frac{B}{x-2}.$$

Multiplying by $(x-1)$ and $(x-2)$, this becomes

$$3x+1 = A(x-2) + B(x-1)$$
$$= (A+B)x - 2A - B$$

which produces the system of equations

$$\begin{cases} A+B = 3 \\ -2A - B = 1. \end{cases}$$

Solving this system yields $A = -4$ and $B = 7$. So,

$$\int \frac{3x+1}{x^2-3x+2} dx = \int \left(-\frac{4}{x-1} + \frac{7}{x-2} \right) dx$$
$$= -4 \int \frac{dx}{x-1} + 7 \int \frac{dx}{x-2}$$
$$= -4 \ln|x-1| + 7 \ln|x-2| + C.$$

129. To find $\int we^{-w} \, dw$, integrate by parts, with $u = w$ and $v' = e^{-w}$. Then $u' = 1$ and $v = -e^{-w}$. Then

$$\int we^{-w} \, dw = -we^{-w} + \int e^{-w} \, dw = -we^{-w} - e^{-w} + C.$$

Thus

$$\int_0^\infty we^{-w} \, dw = \lim_{b \to \infty} \int_0^b we^{-w} \, dw = \lim_{b \to \infty} (-we^{-w} - e^{-w}) \Big|_0^b = 1.$$

133. We find the exact value:

$$\int_{10}^\infty \frac{1}{z^2-4} dz = \int_{10}^\infty \frac{1}{(z+2)(z-2)} dz$$
$$= \lim_{b \to \infty} \int_{10}^b \frac{1}{(z+2)(z-2)} dz$$
$$= \lim_{b \to \infty} \frac{1}{4} (\ln|z-2| - \ln|z+2|) \Big|_{10}^b$$
$$= \frac{1}{4} \lim_{b \to \infty} [(\ln|b-2| - \ln|b+2|) - (\ln 8 - \ln 12)]$$
$$= \frac{1}{4} \lim_{b \to \infty} \left[\left(\ln \frac{b-2}{b+2} \right) + \ln \frac{3}{2} \right]$$
$$= \frac{1}{4} (\ln 1 + \ln 3/2) = \frac{\ln 3/2}{4}.$$

137. The integrand $\frac{x}{x+1} \to 1$ as $x \to \infty$, so there's no way $\int_1^\infty \frac{x}{x+1} \, dx$ can converge.

Problems

141. Since the definition of f is different on $0 \le t \le 1$ than it is on $1 \le t \le 2$, break the definite integral at $t = 1$.

$$\int_0^2 f(t) \, dt = \int_0^1 f(t) \, dt + \int_1^2 f(t) \, dt$$

$$= \int_0^1 t^2\, dt + \int_1^2 (2-t)\, dt$$

$$= \frac{t^3}{3}\bigg|_0^1 + \left(2t - \frac{t^2}{2}\right)\bigg|_1^2$$

$$= 1/3 + 1/2 = 5/6 \approx 0.833$$

145. (a) i. 0   ii. $\frac{2}{\pi}$ iii. $\frac{1}{2}$

  (b) Average value of $f(t) <$ Average value of $k(t) <$ Average value of $g(t)$

We can look at the three functions in the range $-\frac{\pi}{2} \le x \le \frac{3\pi}{2}$, since they all have periods of 2π ($|\cos t|$ and $(\cos t)^2$ also have a period of π, but that doesn't hurt our calculation). It is clear from the graphs of the three functions below that the average value for $\cos t$ is 0 (since the area above the x-axis is equal to the area below it), while the average values for the other two are positive (since they are everywhere positive, except where they are 0).

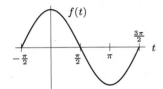 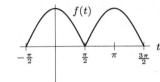 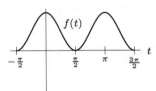

It is also fairly clear from the graphs that the average value of $g(t)$ is greater than the average value of $k(t)$; it is also possible to see this algebraically, since

$$(\cos t)^2 = |\cos t|^2 \le |\cos t|$$

because $|\cos t| \le 1$ (and both of these $\le$'s are $<$'s at all the points where the functions are not 0 or 1).

149. (a) Since the rate is given by $r(t) = 2te^{-2t}$ ml/sec, by the Fundamental Theorem of Calculus, the total quantity is given by the definite integral:

$$\text{Total quantity} \approx \int_0^\infty 2te^{-2t}\, dt = 2\lim_{b\to\infty} \int_0^b te^{-2t}\, dt.$$

Integration by parts with $u = t$, $v' = e^{-2t}$ gives

$$\text{Total quantity} \approx 2\lim_{b\to\infty} \left(-\frac{t}{2}e^{-2t} - \frac{1}{4}e^{-2t}\right)\bigg|_0^b$$

$$= 2\lim_{b\to\infty}\left(\frac{1}{4} - \left(\frac{b}{2} + \frac{1}{4}\right)e^{-2b}\right) = 2 \cdot \frac{1}{4} = 0.5 \text{ ml}.$$

  (b) At the end of 5 seconds,

$$\text{Quantity received} = \int_0^5 2te^{-2t}\, dt \approx 0.49975 \text{ ml}.$$

Since $0.49975/0.5 = 0.9995 = 99.95\%$, the patient has received 99.95% of the dose in the first 5 seconds.

CAS Challenge Problems

153. (a) A possible answer from the CAS is

$$\int \sin^3 x\, dx = \frac{-9\cos(x) + \cos(3\,x)}{12}.$$

  (b) Differentiating

$$\frac{d}{dx}\left(\frac{-9\cos(x) + \cos(3\,x)}{12}\right) = \frac{9\sin(x) - 3\sin(3\,x)}{12} = \frac{3\sin x - \sin(3x)}{4}.$$

(c) Using the identities, we get

$$\sin(3x) = \sin(x + 2x) = \sin x \cos 2x + \cos x \sin 2x$$
$$= \sin x(1 - 2\sin^2 x) + \cos x(2\sin x \cos x)$$
$$= \sin x - 2\sin^3 x + 2\sin x(1 - \sin^2 x)$$
$$= 3\sin x - 4\sin^3 x.$$

Thus,

$$3\sin x - \sin(3x) = 3\sin x - (3\sin x - 4\sin^3 x) = 4\sin^3 x,$$

so

$$\frac{3\sin x - \sin(3x)}{4} = \sin^3 x.$$

CHECK YOUR UNDERSTANDING

1. False. The subdivision size $\Delta x = (1/10)(6 - 2) = 4/10$.

5. True. We have

$$\text{LEFT}(n) - \text{RIGHT}(n) = (f(x_0) + f(x_1) + \cdots + f(x_{n-1}))\Delta x - (f(x_1) + f(x_2) + \cdots + f(x_n))\Delta x.$$

On the right side of the equation, all terms cancel except the first and last, so:

$$\text{LEFT}(n) - \text{RIGHT}(n) = (f(x_0) - f(x_n))\Delta x = (f(2) - f(6))\Delta x.$$

This is also discussed in Section 5.1.

9. False. This is true if f is an increasing function or if f is a decreasing function, but it is not true in general. For example, suppose that $f(2) = f(6)$. Then $\text{LEFT}(n) = \text{RIGHT}(n)$ for all n, which means that if $\int_2^6 f(x)dx$ lies between $\text{LEFT}(n)$ and $\text{RIGHT}(n)$, then it must equal $\text{LEFT}(n)$, which is not always the case.
   For example, if $f(x) = (x - 4)^2$ and $n = 1$, then $f(2) = f(6) = 4$, so

$$\text{LEFT}(1) = \text{RIGHT}(1) = 4 \cdot (6 - 2) = 16.$$

However

$$\int_2^6 (x - 4)^2 dx = \frac{(x - 4)^3}{3}\Big|_2^6 = \frac{2^3}{3} - \left(-\frac{2^3}{3}\right) = \frac{16}{3}.$$

In this example, since $\text{LEFT}(n) = \text{RIGHT}(n)$, we have $\text{TRAP}(n) = \text{LEFT}(n)$. However trapezoids overestimate the area, since the graph of f is concave up. This is also discussed in Section 7.5.

13. True. Rewrite $\sin^7 \theta = \sin \theta \sin^6 \theta = \sin \theta(1 - \cos^2 \theta)^3$. Expanding, substituting $w = \cos \theta, dw = -\sin \theta \, d\theta$, and integrating gives a polynomial in w, which is a polynomial in $\cos \theta$.

17. True. Let $u = t, v' = \sin(5 - t)$, so $u' = 1, v = \cos(5 - t)$. Then the integral $\int 1 \cdot \cos(5 - t) \, dt$ can be done by guess-and-check or by substituting $w = 5 - t$.

21. False. Let $f(x) = x + 1$. Then

$$\int_0^\infty \frac{1}{x + 1} dx = \lim_{b \to \infty} \ln|x + 1|\Big|_0^b = \lim_{b \to \infty} \ln(b + 1),$$

but $\lim_{b \to \infty} \ln(b + 1)$ does not exist.

25. True. Make the substitution $w = ax$. Then $dw = a \, dx$, so

$$\int_0^b f(ax) \, dx = \frac{1}{a} \int_0^c f(w) \, dw,$$

where $c = ab$. As b approaches infinity, so does c, since a is constant. Thus the limit of the left side of the equation as b approaches infinity is finite exactly when the limit of the right side of the equation as c approaches infinity is finite. That is, $\int_0^\infty f(ax) \, dx$ converges exactly when $\int_0^\infty f(x) \, dx$ converges.

CHAPTER EIGHT

Solutions for Section 8.1

Exercises

1. Each strip is a rectangle of length 3 and width Δx, so

$$\text{Area of strip } = 3\Delta x, \quad \text{so}$$

$$\text{Area of region } = \int_0^5 3\,dx = 3x \Big|_0^5 = 15.$$

Check: This area can also be computed using Length $\times$ Width $= 5 \cdot 3 = 15$.

5. The strip has width Δy, so the variable of integration is y. The length of the strip is x. Since $x^2 + y^2 = 10$ and the region is in the first quadrant, solving for x gives $x = \sqrt{10 - y^2}$. Thus

$$\text{Area of strip } \approx x\Delta y = \sqrt{10 - y^2}\,dy.$$

The region stretches from $y = 0$ to $y = \sqrt{10}$, so

$$\text{Area of region } = \int_0^{\sqrt{10}} \sqrt{10 - y^2}\,dy.$$

Evaluating using VI-30 from the Table of Integrals, we have

$$\text{Area } = \frac{1}{2}\left(y\sqrt{10 - y^2} + 10\arcsin\left(\frac{y}{\sqrt{10}}\right)\right)\Bigg|_0^{\sqrt{10}} = 5(\arcsin 1 - \arcsin 0) = \frac{5}{2}\pi.$$

Check: This area can also be computed using the formula $\frac{1}{4}\pi r^2 = \frac{1}{4}\pi(\sqrt{10})^2 = \frac{5}{2}\pi$.

9. Each slice is a circular disk with radius $r = 2$ cm.

$$\text{Volume of disk } = \pi r^2 \Delta x = 4\pi \Delta x \text{ cm}^3.$$

Summing over all disks, we have

$$\text{Total volume } \approx \sum 4\pi \Delta x \text{ cm}^3.$$

Taking a limit as $\Delta x \to 0$, we get

$$\text{Total volume } = \lim_{\Delta x \to 0} \sum 4\pi \Delta x = \int_0^9 4\pi\,dx \text{ cm}^3.$$

Evaluating gives

$$\text{Total volume } = 4\pi x \Big|_0^9 = 36\pi \text{ cm}^3.$$

Check: The volume of the cylinder can also be calculated using the formula $V = \pi r^2 h = \pi 2^2 \cdot 9 = 36\pi$ cm^3.

13. Each slice is a circular disk. See Figure 8.1. The radius of the sphere is 5 mm, and the radius r at height y is given by the Pythagorean Theorem

$$y^2 + r^2 = 5^2.$$

Solving gives $r = \sqrt{5^2 - y^2}$ mm. Thus,

$$\text{Volume of disk } \approx \pi r^2 \Delta y = \pi(5^2 - y^2)\Delta y \text{ mm}^3.$$

Summing over all disks, we have

$$\text{Total volume} \approx \sum \pi(5^2 - y^2)\Delta y \text{ mm}^3.$$

Taking the limit as $\Delta y \to 0$, we get

$$\text{Total volume} = \lim_{\Delta y \to 0} \sum \pi(5^2 - y^2)\Delta y = \int_0^5 \pi(5^2 - y^2)\, dy \text{ mm}^3.$$

Evaluating gives

$$\text{Total volume} = \pi \left(25y - \frac{y^3}{3} \right)\Bigg|_0^5 = \frac{250}{3}\pi \text{ mm}^3.$$

Check: The volume of a hemisphere can be calculated using the formula $V = \frac{2}{3}\pi r^3 = \frac{2}{3}\pi 5^3 = \frac{250}{3}\pi \text{ mm}^3$.

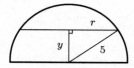

Figure 8.1

Problems

17. Triangle of base and height 1 and 3. See Figure 8.2. (Either 1 or 3 can be the base. A non-right triangle is also possible.)

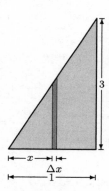

Figure 8.2

21. Hemisphere with radius 12. See Figure 8.3.

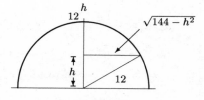

Figure 8.3

25.

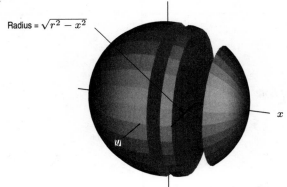

Radius = $\sqrt{r^2 - x^2}$

We slice up the sphere in planes perpendicular to the x-axis. Each slice is a circle, with radius $y = \sqrt{r^2 - x^2}$; that's the radius because $x^2 + y^2 = r^2$ when $z = 0$. Then the volume is

$$V \approx \sum \pi(y^2)\,\Delta x = \sum \pi(r^2 - x^2)\,\Delta x.$$

Therefore, as Δx tends to zero, we get

$$V = \int_{x=-r}^{x=r} \pi(r^2 - x^2)\,dx$$
$$= 2\int_{x=0}^{x=r} \pi(r^2 - x^2)\,dx$$
$$= 2\left(\pi r^2 x - \frac{\pi x^3}{3}\right)\Big|_0^r$$
$$= \frac{4\pi r^3}{3}.$$

29. To calculate the volume of material, we slice the dam horizontally. See Figure 8.4. The slices are rectangular, so

$$\text{Volume of slice} \approx 1400 w \Delta h \text{ m}^3.$$

Since w is a linear function of h, and $w = 160$ when $h = 0$, and $w = 10$ when $h = 150$, this function has slope $= (10 - 160)/150 = -1$. Thus

$$w = 160 - h \text{ meters},$$

so

$$\text{Volume of slice} \approx 1400(160 - h)\Delta h \text{ m}^3.$$

Summing over all slices and taking the limit as $\Delta h \to 0$ gives

$$\text{Total volume} = \lim_{\Delta h \to 0} \sum 1400(160 - h)\Delta h = \int_0^{150} 1400(160 - h)\,dh \text{ m}^3.$$

Evaluating the integral gives

$$\text{Total volume} = 1400\left(160h - \frac{h^2}{2}\right)\Big|_0^{150} = 1.785 \cdot 10^7 \text{ m}^3.$$

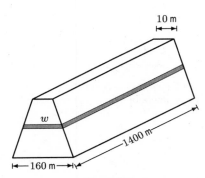

Figure 8.4

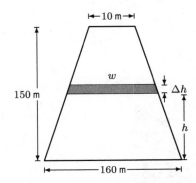

Figure 8.5

Solutions for Section 8.2

Exercises

1. The volume is given by

$$V = \int_0^1 \pi y^2 \, dx = \int_0^1 \pi x^4 \, dx = \pi \frac{x^5}{5} \bigg|_0^1 = \frac{\pi}{5}.$$

5. The volume is given by

$$V = \int_{-2}^0 \pi (4 - x^2)^2 \, dx = \pi \int_{-2}^0 (16 - 8x^2 + x^4) \, dx = \pi \left(16x - \frac{8x^3}{3} + \frac{x^5}{5} \right) \bigg|_{-2}^0 = \frac{256\pi}{15}.$$

9. Since the graph of $y = e^{3x}$ is above the graph of $y = e^x$ for $0 \le x \le 1$, the volume is given by

$$V = \int_0^1 \pi (e^{3x})^2 \, dx - \int_0^1 \pi (e^x)^2 \, dx = \int_0^1 \pi (e^{6x} - e^{2x}) \, dx = \pi \left(\frac{e^{6x}}{6} - \frac{e^{2x}}{2} \right) \bigg|_0^1 = \pi \left(\frac{e^6}{6} - \frac{e^2}{2} + \frac{1}{3} \right).$$

13. We have

$$D = \int_0^1 \sqrt{(-e^t \sin(e^t))^2 + (e^t \cos(e^t))^2} \, dt$$

$$= \int_0^1 \sqrt{e^{2t}} \, dt = \int_0^1 e^t \, dt$$

$$= e - 1.$$

This is the length of the arc of a unit circle from the point $(\cos 1, \sin 1)$ to $(\cos e, \sin e)$—in other words between the angles $\theta = 1$ and $\theta = e$. The length of this arc is $(e - 1)$.

Problems

17.

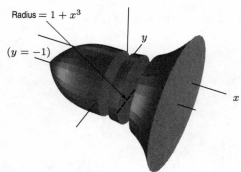

Radius $= 1 + x^3$

$(y = -1)$

We slice the region perpendicular to the x–axis. The Riemann sum we get is $\sum \pi (x^3 + 1)^2 \Delta x$. So the volume V is the integral

$$V = \int_{-1}^1 \pi (x^3 + 1)^2 \, dx$$

$$= \pi \int_{-1}^1 (x^6 + 2x^3 + 1) \, dx$$

$$= \pi \left(\frac{x^7}{7} + \frac{x^4}{2} + x \right) \bigg|_{-1}^1$$

$$= (16/7)\pi \approx 7.18.$$

21.

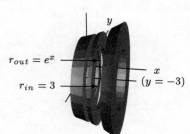

$r_{out} = e^x$

$r_{in} = 3$

$(y = -3)$

We slice the volume with planes perpendicular to the line $y = -3$. This divides the curve into thin washers, as in Example 3 on page 354 of the text, whose volumes are

$$\pi r_{out}^2 dx - \pi r_{in}^2 dx = \pi (3 + y)^2 dx - \pi 3^2 dx.$$

So the integral we get from adding all these washers up is

$$V = \int_{x=0}^{x=1} [\pi (3 + y)^2 - \pi 3^2] \, dx$$

$$= \pi \int_0^1 [(3 + e^x)^2 - 9] \, dx$$

$$= \pi \int_0^1 [e^{2x} + 6e^x] \, dx = \pi [\frac{e^{2x}}{2} + 6e^x] \bigg|_0^1$$

$$= \pi [(e^2/2 + 6e) - (1/2 + 6)] \approx 42.42.$$

25. (a) We can begin by slicing the pie into horizontal slabs of thickness Δh located at height h. To find the radius of each slice, we note that radius increases linearly with height. Since $r = 4.5$ when $h = 3$ and $r = 3.5$ when $h = 0$, we should have $r = 3.5 + h/3$. Then the volume of each slab will be $\pi r^2 \, \Delta h = \pi(3.5 + h/3)^2 \, \Delta h$. To find the total volume of the pie, we integrate this from $h = 0$ to $h = 3$:

$$V = \pi \int_0^3 \left(3.5 + \frac{h}{3}\right)^2 \, dh$$

$$= \pi \left[\frac{h^3}{27} + \frac{7h^2}{6} + \frac{49h}{4}\right]\Big|_0^3$$

$$= \pi \left[\frac{3^3}{27} + \frac{7(3^2)}{6} + \frac{49(3)}{4}\right] \approx 152 \text{ in}^3.$$

(b) We use 1.5 in as a rough estimate of the radius of an apple. This gives us a volume of $(4/3)\pi(1.5)^3 \approx 10 \text{ in}^3$. Since $152/10 \approx 15$, we would need about 15 apples to make a pie.

29. (a) The volume, V, contained in the bowl when the surface has height h is

$$V = \int_0^h \pi x^2 \, dy.$$

However, since $y = x^4$, we have $x^2 = \sqrt{y}$ so that

$$V = \int_0^h \pi \sqrt{y} \, dy = \frac{2}{3}\pi h^{3/2}.$$

Differentiating gives $dV/dh = \pi h^{1/2} = \pi\sqrt{h}$. We are given that $dV/dt = -6\sqrt{h}$, where the negative sign reflects the fact that V is decreasing. Using the chain rule we have

$$\frac{dh}{dt} = \frac{dh}{dV} \cdot \frac{dV}{dt} = \frac{1}{dV/dh} \cdot \frac{dV}{dt} = \frac{1}{\pi\sqrt{h}} \cdot (-6\sqrt{h}) = -\frac{6}{\pi}.$$

Thus, $dh/dt = -6/\pi$, a constant.

(b) Since $dh/dt = -6/\pi$ we know that $h = -6t/\pi + C$. However, when $t = 0$, $h = 1$, therefore $h = 1 - 6t/\pi$. The bowl is empty when $h = 0$, that is when $t = \pi/6$ units.

33. Since $y = (e^x + e^{-x})/2$, $y' = (e^x - e^{-x})/2$. The length of the catenary is

$$\int_{-1}^1 \sqrt{1 + (y')^2} \, dx = \int_{-1}^1 \sqrt{1 + \left[\frac{e^x - e^{-x}}{2}\right]^2} \, dx = \int_{-1}^1 \sqrt{1 + \frac{e^{2x}}{4} - \frac{1}{2} + \frac{e^{-2x}}{4}} \, dx$$

$$= \int_{-1}^1 \sqrt{\left[\frac{e^x + e^{-x}}{2}\right]^2} \, dx = \int_{-1}^1 \frac{e^x + e^{-x}}{2} \, dx$$

$$= \left[\frac{e^x - e^{-x}}{2}\right]\Big|_{-1}^1 = e - e^{-1}.$$

Solutions for Section 8.3

Exercises

1. Since density is e^{-x} gm/cm,

$$\text{Mass} = \int_0^{10} e^{-x} \, dx = -e^{-x}\Big|_0^{10} = 1 - e^{-10} \text{ gm}.$$

5. (a) Figure 8.6 shows a graph of the density function.

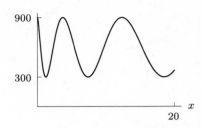

Figure 8.6

(b) Suppose we choose an x, $0 \leq x \leq 20$. We approximate the density of the number of the cars between x and $x + \Delta x$ miles as $\delta(x)$ cars per mile. Therefore, the number of cars between x and $x + \Delta x$ is approximately $\delta(x)\Delta x$. If we slice the 20 mile strip into N slices, we get that the total number of cars is

$$C \approx \sum_{i=1}^{N} \delta(x_i)\Delta x = \sum_{i=1}^{N} \left[600 + 300 \sin(4\sqrt{x_i + 0.15}) \right] \Delta x,$$

where $\Delta x = 20/N$. (This is a right-hand approximation; the corresponding left-hand approximation is $\displaystyle\sum_{i=0}^{N-1} \delta(x_i)\Delta x$.)

(c) As $N \to \infty$, the Riemann sum above approaches the integral

$$C = \int_{0}^{20} (600 + 300 \sin 4\sqrt{x + 0.15})\, dx.$$

If we calculate the integral numerically, we find $C \approx 11513$. We can also find the integral exactly as follows:

$$C = \int_{0}^{20} (600 + 300 \sin 4\sqrt{x + 0.15})\, dx$$

$$= \int_{0}^{20} 600\, dx + \int_{0}^{20} 300 \sin 4\sqrt{x + 0.15}\, dx$$

$$= 12000 + 300 \int_{0}^{20} \sin 4\sqrt{x + 0.15}\, dx.$$

Let $w = \sqrt{x + 0.15}$, so $x = w^2 - 0.15$ and $dx = 2w\, dw$. Then

$$\int_{x=0}^{x=20} \sin 4\sqrt{x + 0.15}\, dx = 2 \int_{w=\sqrt{0.15}}^{w=\sqrt{20.15}} w \sin 4w\, dw, \text{ (using integral table III-15)}$$

$$= 2 \left[-\frac{1}{4} w \cos 4w + \frac{1}{16} \sin 4w \right] \Bigg|_{\sqrt{0.15}}^{\sqrt{20.15}}$$

$$\approx -1.624.$$

Using this, we have $C \approx 12000 + 300(-1.624) \approx 11513$, which matches our numerical approximation.

Problems

9. Since the density varies with x, the region must be sliced perpendicular to the x-axis. This has the effect of making the density approximately constant on each strip. See Figure 8.7. Since a strip is of height y, its area is approximately $y\Delta x$. The density on the strip is $\delta(x) = 1 + x$ gm/cm^2. Thus

$$\text{Mass of strip} \approx \text{Density} \cdot \text{Area} \approx (1 + x)y\Delta x \text{ gm.}$$

Because the tops of the strips end on two different lines, one for $x \geq 0$ and the other for $x < 0$, the mass is calculated as the sum of two integrals. See Figure 8.7. For the left part of the region, $y = x + 1$, so

$$\text{Mass of left part} = \lim_{\Delta x \to 0} \sum (1+x)y\Delta x = \int_{-1}^{0} (1+x)(x+1)\,dx$$

$$= \int_{-1}^{0} (1+x)^2\,dx = \left.\frac{(x+1)^3}{3}\right|_{-1}^{0} = \frac{1}{3}\text{ gm.}$$

From Figure 8.7, we see that for the right part of the region, $y = -x + 1$, so

$$\text{Mass of right part} = \lim_{\Delta x \to 0} \sum (1+x)y\Delta x = \int_{0}^{1} (1+x)(-x+1)\,dx$$

$$= \int_{0}^{1} (1-x^2)\,dx = \left. x - \frac{x^3}{3}\right|_{0}^{1} = \frac{2}{3}\text{ gm.}$$

$$\text{Total mass} = \frac{1}{3} + \frac{2}{3} = 1\text{ gm.}$$

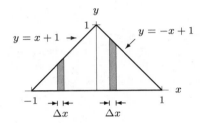

Figure 8.7

13. **(a)** Use the formula for the volume of a cylinder:

$$\text{Volume} = \pi r^2 l.$$

Since it is only a half cylinder

$$\text{Volume of shed} = \frac{1}{2}\pi r^2 l.$$

(b) Set up the axes as shown in Figure 8.8. The density can be defined as

$$\text{Density} = ky.$$

Now slice the sawdust horizontally into slabs of thickness Δy as shown in Figure 8.9, and calculate

$$\text{Volume of slab} \approx 2xl\Delta y = 2l(\sqrt{r^2 - y^2})\Delta y.$$

$$\text{Mass of slab} = \text{Density} \cdot \text{Volume} \approx 2kly\sqrt{r^2 - y^2}\,\Delta y.$$

Finally, we compute the total mass of sawdust:

$$\text{Total mass of sawdust} = \int_{0}^{r} 2kly\sqrt{r^2 - y^2}\,dy = \left. -\frac{2}{3}kl(r^2 - y^2)^{3/2}\right|_{0}^{r} = \frac{2klr^3}{3}.$$

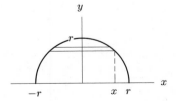

Figure 8.8

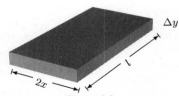

Figure 8.9

17. We need the numerator of $\bar{x}$, to be zero, i.e. $\sum x_i m_i = 0$. Since all of the masses are the same, we can factor them out and write $4 \sum x_i = 0$. Thus the fourth mass needs to be placed so that all of the positions sum to zero. The first three positions sum to $(-6 + 1 + 3) = -2$, so the fourth mass needs to be placed at $x = 2$.

21. (a) The density is minimum at $x = -1$ and increases as x increases, so more of the mass of the rod is in the right half of the rod. We thus expect the balancing point to be to the right of the origin.

(b) We need to compute

$$\int_{-1}^{1} x(3 - e^{-x})\, dx = \left(\frac{3}{2}x^2 + xe^{-x} + e^{-x}\right)\Big|_{-1}^{1} \quad \text{(using integration by parts)}$$

$$= \frac{3}{2} + e^{-1} + e^{-1} - \left(\frac{3}{2} - e^1 + e^1\right) = \frac{2}{e}.$$

We must divide this result by the total mass, which is given by

$$\int_{-1}^{1} (3 - e^{-x})\, dx = (3x + e^{-x})\Big|_{-1}^{1} = 6 - e + \frac{1}{e}.$$

We therefore have

$$\bar{x} = \frac{2/e}{6 - e + (1/e)} = \frac{2}{1 + 6e - e^2} \approx 0.2.$$

25. Stand the cone with the base horizontal, with center at the origin. Symmetry gives us that $\bar{x} = \bar{y} = 0$. Since the cone is fatter near its base we expect the center of mass to be nearer to the base.

Slice the cone into disks parallel to the xy-plane.

As we saw in Example 2 on page 347, a disk of thickness Δz at height z above the base has

$$\text{Volume of disk} = A_z(z)\Delta z \approx \pi(5 - z)^2 \Delta z \text{ cm}^3.$$

Thus, since the density is δ,

$$\bar{z} = \frac{\int z\delta A_z(z)\, dz}{\text{Mass}} = \frac{\int_0^5 z \cdot \delta\pi(5 - z)^2\, dz}{\text{Mass}} \text{ cm}.$$

To evaluate the integral in the numerator, we factor out the constant density δ and π to get

$$\int_0^5 z \cdot \delta\pi(5 - z)^2\, dz = \delta\pi \int_0^5 z(25 - 10z + z^2)\, dz = \delta\pi \left(\frac{25z^2}{2} - \frac{10z^3}{3} + \frac{z^4}{4}\right)\Big|_0^5 = \frac{625}{12}\delta\pi.$$

We divide this result by the total mass of the cone, which is $\left(\frac{1}{3}\pi 5^2 \cdot 5\right)\delta$:

$$\bar{z} = \frac{\frac{625}{12}\delta\pi}{\frac{1}{3}\pi 5^3 \delta} = \frac{5}{4} = 1.25 \text{ cm}.$$

As predicted, the center of mass is closer to the base of the cone than its top.

Solutions for Section 8.4

Exercises

1. The work done is given by

$$W = \int_1^2 3x\, dx = \frac{3}{2}x^2 \Big|_1^2 = \frac{9}{2} \text{ joules}.$$

5. The force exerted on the satellite by the earth (and vice versa!) is GMm/r^2, where r is the distance from the center of the earth to the center of the satellite, m is the mass of the satellite, M is the mass of the earth, and G is the gravitational constant. So the total work done is

$$\int_{6.4\cdot10^6}^{8.4\cdot10^6} F\, dr = \int_{6.4\cdot10^6}^{8.4\cdot10^6} \frac{GMm}{r^2}\, dr = \left(\frac{-GMm}{r}\right)\bigg|_{6.4\cdot10^6}^{8.4\cdot10^6} \approx 1.489\cdot10^{10}\text{ joules.}$$

Problems

9. Consider lifting a rectangular slab of water h feet from the top up to the top. The area of such a slab is $(10)(20) = 200$ square feet; if the thickness is dh, then the volume of such a slab is $200\, dh$ cubic feet. This much water weighs 62.4 pounds per ft^3, so the weight of such a slab is $(200\, dh)(62.4) = 12480\, dh$ pounds. To lift that much water h feet requires $12480h\, dh$ foot-pounds of work. To lift the whole tank, we lift one plate at a time; integrating over the slabs yields

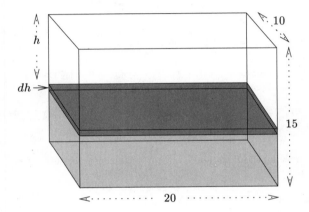

$$\int_0^{15} 12480h\, dh = \frac{12480h^2}{2}\bigg|_0^{15} = \frac{12480\cdot15^2}{2} = 1{,}404{,}000\text{ foot-pounds.}$$

13.

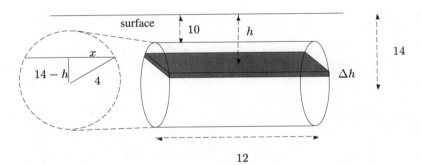

Let h represent distance below the surface in feet. We slice the tank up into horizontal slabs of thickness Δh. From looking at the figure, we can see that the slabs will be rectangular. The length of any slab is 12 feet. The width w of a slab h units below the ground will equal $2x$, where $(14 - h)^2 + x^2 = 16$, so $w = 2\sqrt{4^2 - (14 - h)^2}$. The volume of such a slab is therefore $12w\, \Delta h = 24\sqrt{16 - (14 - h)^2}\, \Delta h$ cubic feet; the slab weighs $42\cdot24\sqrt{16 - (14 - h)^2}\, \Delta h = 1008\sqrt{16 - (14 - h)^2}\, \Delta h$ pounds. So the total work done in pumping out all the gasoline is

$$\int_{10}^{18} 1008h\sqrt{16 - (14 - h)^2}\, dh = 1008\int_{10}^{18} h\sqrt{16 - (14 - h)^2}\, dh.$$

Substitute $s = 14 - h$, $ds = -dh$. We get

$$1008 \int_{10}^{18} h\sqrt{16 - (14 - h)^2}\, dh = -1008 \int_{4}^{-4} (14 - s)\sqrt{16 - s^2}\, ds$$

$$= 1008 \cdot 14 \int_{-4}^{4} \sqrt{16 - s^2}\, ds - 1008 \int_{-4}^{4} s\sqrt{16 - s^2}\, ds.$$

The first integral represents the area of a semicircle of radius 4, which is 8π. The second is the integral of an odd function, over the interval $-4 \le s \le 4$, and is therefore 0. Hence, the total work is $1008 \cdot 14 \cdot 8\pi \approx 354{,}673$ foot-pounds.

17. Bottom:

$$\text{Water force } = 62.4(2)(12) = 1497.6 \text{ lbs.}$$

Front and back:

$$\text{Water force } = (62.4)(4) \int_{0}^{2} (2 - x)\, dx = (62.4)(4)\left(2x - \frac{1}{2}x^2\right)\Big|_{0}^{2}$$

$$= (62.4)(4)(2) = 499.2 \text{ lbs.}$$

Both sides:

$$\text{Water force } = (62.4)(3) \int_{0}^{2} (2 - x)\, dx = (62.4)(3)(2) = 374.4 \text{ lbs.}$$

21. We divide the dam into horizontal strips since the pressure is then approximately constant on each one. See Figure 8.10.

$$\text{Area of strip } \approx w\Delta h \text{ m}^2.$$

Since w is a linear function of h, and $w = 3600$ when $h = 0$, and $w = 3000$ when $h = 100$, the function has slope $(3000 - 3600)/100 = -6$. Thus,

$$w = 3600 - 6h,$$

so

$$\text{Area of strip } \approx (3600 - 6h)\Delta h \text{ m}^2.$$

The density of water is $\delta = 1000 \text{ kg/m}^3$, so the pressure at depth h meters $= \delta g h = 1000 \cdot 9.8h = 9800h \text{ nt/m}^2$. Thus,

$$\text{Total force } = \lim_{\Delta h \to 0} \sum 9800h(3600 - 6h)\Delta h = 9800 \int_{0}^{100} h(3600 - 6h)\, dh \text{ newtons.}$$

Evaluating the integral gives

$$\text{Total force } = 9800(1800h^2 - 2h^3)\Big|_{0}^{100} = 1.6 \cdot 10^{11} \text{ newtons.}$$

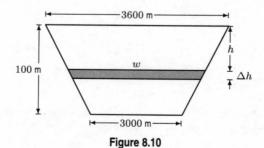

Figure 8.10

25.

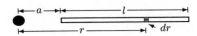

The density of the rod, in mass per unit length, is M/l (see above). So a slice of size dr has mass $\frac{M\,dr}{l}$. It pulls the small mass m with force $Gm\frac{M\,dr}{l}/r^2 = \frac{GmM\,dr}{lr^2}$. So the total gravitational attraction between the rod and point is

$$\int_a^{a+l} \frac{GmM\,dr}{lr^2} = \frac{GmM}{l}\left(-\frac{1}{r}\right)\Bigg|_a^{a+l}$$

$$= \frac{GmM}{l}\left(\frac{1}{a} - \frac{1}{a+l}\right)$$

$$= \frac{GmM}{l}\frac{l}{a(a+l)} = \frac{GmM}{a(a+l)}.$$

Solutions for Section 8.5

Exercises

1. At any time t, in a time interval Δt, an amount of $1000\Delta t$ is deposited into the account. This amount earns interest for $(10 - t)$ years giving a future value of $1000e^{(0.08)(10-t)}$. Summing all such deposits, we have

$$\text{Future value} = \int_0^{10} 1000e^{0.08(10-t)}\,dt = \$15{,}319.30.$$

Problems

5.

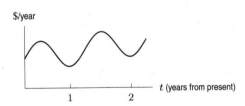

The graph reaches a peak each summer, and a trough each winter. The graph shows sunscreen sales increasing from cycle to cycle. This gradual increase may be due in part to inflation and to population growth.

9. You should choose the payment which gives you the highest present value. The immediate lump-sum payment of \$2800 obviously has a present value of exactly \$2800, since you are getting it now. We can calculate the present value of the installment plan as:

$$\text{PV} = 1000e^{-0.06(0)} + 1000e^{-0.06(1)} + 1000e^{-0.06(2)}$$

$$\approx \$2828.68.$$

Since the installment payments offer a (slightly) higher present value, you should accept this option.

13. (a) Suppose the oil extracted over the time period $[0, M]$ is S. (See Figure 8.11.) Since $q(t)$ is the rate of oil extraction, we have:

$$S = \int_0^M q(t)dt = \int_0^M (a - bt)dt = \int_0^M (10 - 0.1t)\,dt.$$

To calculate the time at which the oil is exhausted, set $S = 100$ and try different values of M. We find $M = 10.6$ gives

$$\int_0^{10.6} (10 - 0.1t)\,dt = 100,$$

so the oil is exhausted in 10.6 years.

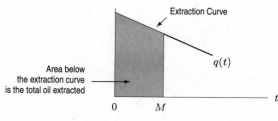

Figure 8.11

(b) Suppose p is the oil price, C is the extraction cost per barrel, and r is the interest rate. We have the present value of the profit as

$$\text{Present value of profit} = \int_0^M (p - C)q(t)e^{-rt}dt$$

$$= \int_0^{10.6} (20 - 10)(10 - 0.1t)e^{-0.1t}\, dt$$

$$= 624.9 \text{ million dollars.}$$

17.

$$\int_0^{q^*} (p^* - S(q))\, dq = \int_0^{q^*} p^*\, dq - \int_0^{q^*} S(q)\, dq$$

$$= p^*q^* - \int_0^{q^*} S(q)\, dq.$$

Using Problem 16, this integral is the extra amount consumers pay (i.e., suppliers earn over and above the minimum they would be willing to accept for supplying the good). It results from charging the equilibrium price.

Solutions for Section 8.6

Exercises

1.

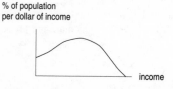

Figure 8.12: Density function

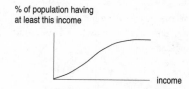

Figure 8.13: Cumulative distribution function

5. Since the function is decreasing, it cannot be a cdf (whose values never decrease). Thus, the function is a pdf.
   The area under a pdf is 1, so, using the formula for the area of a triangle, we have

$$\frac{1}{2}4c = 1, \quad \text{giving} \quad c = \frac{1}{2}.$$

The pdf is

$$p(x) = \frac{1}{2} - \frac{1}{8}x \quad \text{for} \quad 0 \le x \le 4,$$

so the cdf is given in Figure 8.14 by

$$P(x) = \begin{cases} 0 & \text{for} \quad x < 0 \\ \dfrac{x}{2} - \dfrac{x^2}{16} & \text{for} \quad 0 \le x \le 4 \\ 1 & \text{for} \quad x > 4. \end{cases}$$

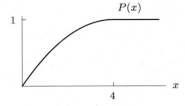

Figure 8.14

9. This function does not level off to 1, and it is not always increasing. Thus, the function is a pdf. Since the area under the curve must be 1, using the formula for the area of a triangle,

$$\frac{1}{2} \cdot c \cdot 1 = 1 \quad \text{so} \quad c = 2.$$

Thus, the pdf is given by

$$p(x) = \begin{cases} 0 & \text{for} \quad x < 0 \\ 4x & \text{for} \quad 0 \le x \le 0.5 \\ 2 - 4(x - 0.5) = 4 - 4x & \text{for} \quad 0.5 < x \le 1 \\ 0 & \text{for} \quad x > 0. \end{cases}$$

To find the cdf, we integrate each part of the function separately, making sure that the constants of integration are arranged so that the cdf is continuous.

Since $\int 4x\,dx = 2x^2 + C$ and $P(0) = 0$, we have $2(0)^2 + C = 0$ so $C = 0$. Thus $P(x) = 2x^2$ on $0 \le x \le 0.5$. At $x = 0.5$, the cdf has value $P(0.5) = 2(0.5)^2 = 0.5$. Thus, we arrange that the integral of $4 - 4x$ goes through the point $(0.5, 0.5)$. Since $\int (4 - 4x)\,dx = 4x - 2x^2 + C$, we have

$$4(0.5) - 2(0.5)^2 + C = 0.5 \quad \text{giving} \quad C = -1.$$

Thus

$$P(x) = \begin{cases} 0 & \text{for} \quad x < 0 \\ 2x^2 & \text{for} \quad 0 \le x \le 0.5 \\ 4x - 2x^2 - 1 & \text{for} \quad 0.5 < x \le 1 \\ 1 & \text{for} \quad x > 1. \end{cases}$$

See Figure 8.15.

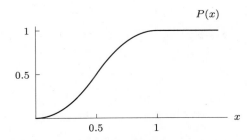

Figure 8.15

Problems

13. For a small interval Δx around 68, the fraction of the population of American men with heights in this interval is about $(0.2)\Delta x$. For example, taking $\Delta x = 0.1$, we can say that approximately $(0.2)(0.1) = 0.02 = 2\%$ of American men have heights between 68 and 68.1 inches.

17. (a) The percentage of calls lasting from 1 to 2 minutes is given by the integral

$$\int_1^2 p(x)\,dx \int_1^2 0.4e^{-0.4x}\,dx = e^{-0.4} - e^{-0.8} \approx 22.1\%.$$

(b) A similar calculation (changing the limits of integration) gives the percentage of calls lasting 1 minute or less as

$$\int_0^1 p(x)\,dx = \int_0^1 0.4e^{-0.4x}\,dx = 1 - e^{-0.4} \approx 33.0\%.$$

(c) The percentage of calls lasting 3 minutes or more is given by the improper integral

$$\int_3^\infty p(x)\,dx = \lim_{b\to\infty}\int_3^b 0.4e^{-0.4x}\,dx = \lim_{b\to\infty}\left(e^{-1.2} - e^{-0.4b}\right) = e^{-1.2} \approx 30.1\%.$$

(d) The cumulative distribution function is the integral of the probability density; thus,

$$C(h) = \int_0^h p(x)\,dx = \int_0^h 0.4e^{-0.4x}\,dx = 1 - e^{-0.4h}.$$

Solutions for Section 8.7

Exercises

1.

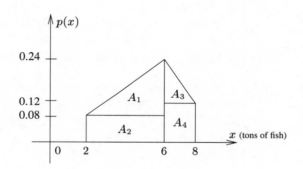

Splitting the figure into four pieces, we see that

$$\text{Area under the curve} = A_1 + A_2 + A_3 + A_4$$
$$= \frac{1}{2}(0.16)4 + 4(0.08) + \frac{1}{2}(0.12)2 + 2(0.12)$$
$$= 1.$$

We expect the area to be 1, since $\int_{-\infty}^{\infty} p(x)\,dx = 1$ for any probability density function, and $p(x)$ is 0 except when $2 \le x \le 8$.

Problems

5. (a) We can find the proportion of students by integrating the density $p(x)$ between $x = 1.5$ and $x = 2$:

$$P(2) - P(1.5) = \int_{1.5}^2 \frac{x^3}{4}\,dx$$
$$= \left.\frac{x^4}{16}\right|_{1.5}^2$$
$$= \frac{(2)^4}{16} - \frac{(1.5)^4}{16} = 0.684,$$

so that the proportion is $0.684 : 1$ or 68.4%.

(b) We find the mean by integrating x times the density over the relevant range:

$$\text{Mean} = \int_0^2 x \left(\frac{x^3}{4}\right) dx$$

$$= \int_0^2 \frac{x^4}{4} dx$$

$$= \frac{x^5}{20} \Big|_0^2$$

$$= \frac{2^5}{20} = 1.6 \text{ hours.}$$

(c) The median will be the time T such that exactly half of the students are finished by time T, or in other words

$$\frac{1}{2} = \int_0^T \frac{x^3}{4} dx$$

$$\frac{1}{2} = \frac{x^4}{16} \Big|_0^T$$

$$\frac{1}{2} = \frac{T^4}{16}$$

$$T = \sqrt[4]{8} = 1.682 \text{ hours.}$$

9. (a) Since $\mu = 100$ and $\sigma = 15$:

$$p(x) = \frac{1}{15\sqrt{2\pi}} e^{-\frac{1}{2}\left(\frac{x-100}{15}\right)^2}.$$

(b) The fraction of the population with IQ scores between 115 and 120 is (integrating numerically)

$$\int_{115}^{120} p(x)\, dx = \int_{115}^{120} \frac{1}{15\sqrt{2\pi}} e^{-\frac{(x-100)^2}{450}} dx$$

$$= \frac{1}{15\sqrt{2\pi}} \int_{115}^{120} e^{-\frac{(x-100)^2}{450}} dx$$

$$\approx 0.067 = 6.7\% \text{ of the population.}$$

13. It is not (a) since a probability density must be a non-negative function; not (c) since the total integral of a probability density must be 1; (b) and (d) are probability density functions, but (d) is not a good model. According to (d), the probability that the next customer comes after 4 minutes is 0. In real life there should be a positive probability of not having a customer in the next 4 minutes. So (b) is the best answer.

Solutions for Chapter 8 Review

Exercises

1. The limits of integration are 0 and b, and the rectangle represents the region under the curve $f(x) = h$ between these limits. Thus,

$$\text{Area of rectangle} = \int_0^b h\, dx = hx \Big|_0^b = hb.$$

5. Each slice is a circular disk. The radius, r, of the disk increases with h and is given in the problem by $r = \sqrt{h}$. Thus

$$\text{Volume of slice} \approx \pi r^2 \Delta h = \pi h \Delta h.$$

Summing over all slices, we have

$$\text{Total volume} \approx \sum \pi h \Delta h.$$

Taking a limit as $\Delta h \to 0$, we get

$$\text{Total volume} = \lim_{\Delta h \to 0} \sum \pi h \Delta h = \int_0^{12} \pi h \, dh.$$

Evaluating gives

$$\text{Total volume} = \pi \frac{h^2}{2} \bigg|_0^{12} = 72\pi.$$

Problems

9. (a)

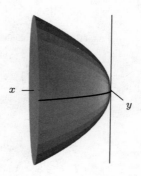

(b) Divide [0,1] into N subintervals of width $\Delta x = \frac{1}{N}$. The volume of the ith disc is $\pi(\sqrt{x_i})^2 \Delta x = \pi x_i \Delta x$. So,
$V \approx \sum_{i=1}^{N} \pi x_i \Delta x$.

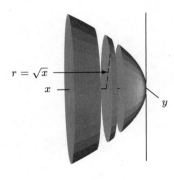

(c)

$$\text{Volume} = \int_0^1 \pi x \, dx = \frac{\pi}{2} x^2 \Big|_0^1 = \frac{\pi}{2} \approx 1.57.$$

13. (a) The line $y = ax$ must pass through (l, b). Hence $b = al$, so $a = b/l$.

(b) Cut the cone into N slices, slicing perpendicular to the x–axis. Each piece is almost a cylinder. The radius of the ith cylinder is $r(x_i) = \dfrac{bx_i}{l}$, so the volume

$$V \approx \sum_{i=1}^{N} \pi \left(\frac{bx_i}{l}\right)^2 \Delta x.$$

Therefore, as $N \to \infty$, we get

$$V = \int_0^l \pi b^2 l^{-2} x^2 dx$$

$$= \pi \frac{b^2}{l^2} \left[\frac{x^3}{3}\right]_0^l = \left(\pi \frac{b^2}{l^2}\right)\left(\frac{l^3}{3}\right) = \frac{1}{3}\pi b^2 l.$$

17. (a) Since the density is constant, the mass is the product of the area of the plate and its density.

$$\text{Area of the plate} = \int_0^1 (\sqrt{x} - x^2)\, dx = \left(\frac{2}{3}x^{3/2} - \frac{1}{3}x^3\right)\Big|_0^1 = \frac{1}{3}\ \text{cm}^2.$$

Thus the mass of the plate is $2 \cdot 1/3 = 2/3$ gm.

(b) See Figure 8.16. Since the region is "fatter" closer to the origin, $\bar{x}$ is less than $1/2$.

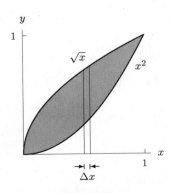

Figure 8.16

(c) To find $\bar{x}$, we slice the region into vertical strips of width Δx. See Figure 8.16.

$$\text{Area of strip} = A_x(x)\Delta x \approx (\sqrt{x} - x^2)\Delta x \text{ cm}^2.$$

Then we have

$$\bar{x} = \frac{\int x\delta A_x(x)\,dx}{\text{Mass}} = \frac{\int_0^1 2x(\sqrt{x} - x^2)\,dx}{2/3} = \frac{3}{2}\int_0^1 2(x^{3/2} - x^3)\,dx = \frac{3}{2}\cdot 2\left(\frac{2}{5}x^{5/2} - \frac{1}{4}x^4\right)\Big|_0^1 = \frac{9}{20} \text{ cm}.$$

This is less than $1/2$, as predicted in part (b). So $\bar{x} = \bar{y} = 9/20$ cm.

21. Let h be height above the bottom of the dam. Then

$$\text{Water force} = \int_0^{25} (62.4)(25 - h)(60)\,dh$$

$$= (62.4)(60)\left(25h - \frac{h^2}{2}\right)\Big|_0^{25}$$

$$= (62.4)(60)(625 - 312.5)$$

$$= (62.4)(60)(312.5)$$

$$= 1{,}170{,}000 \text{ lbs}.$$

25. (a) Slice the mountain horizontally into N cylinders of height Δh. The sum of the volumes of the cylinders will be

$$\sum_{i=1}^{N} \pi r^2 \Delta h = \sum_{i=1}^{N} \pi \left(\frac{3.5\cdot 10^5}{\sqrt{h + 600}}\right)^2 \Delta h.$$

(b)

$$\text{Volume} = \int_{400}^{14400} \pi \left(\frac{3.5\cdot 10^5}{\sqrt{h + 600}}\right)^2 dh$$

$$= 1.23\cdot 10^{11}\pi \int_{400}^{14400} \frac{1}{(h + 600)}\,dh$$

$$= 1.23\cdot 10^{11}\pi \ln(h + 600)\Big|_{400}^{14400} dh$$

$$= 1.23\cdot 10^{11}\pi\,[\ln 15000 - \ln 1000]$$

$$= 1.23\cdot 10^{11}\pi \ln(15000/1000)$$

$$= 1.23\cdot 10^{11}\pi \ln 15 \approx 1.05\cdot 10^{12} \text{ cubic feet}.$$

29. (a) Divide the cross-section of the blood into rings of radius r, width Δr. See Figure 8.17.

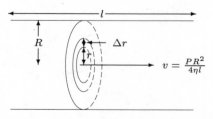

Figure 8.17

Then

$$\text{Area of ring} \approx 2\pi r\Delta r.$$

The velocity of the blood is approximately constant throughout the ring, so

$$\text{Rate blood flows through ring} \approx \text{Velocity} \cdot \text{Area}$$

$$= \frac{P}{4\eta l}(R^2 - r^2) \cdot 2\pi r \Delta r.$$

Thus, summing over all rings, we find the total blood flow:

$$\text{Rate blood flowing through blood vessel} \approx \sum \frac{P}{4\eta l}(R^2 - r^2)2\pi r \Delta r.$$

Taking the limit as $\Delta r \to 0$, we get

$$\text{Rate blood flowing through blood vessel} = \int_0^R \frac{\pi P}{2\eta l}(R^2 r - r^3)dr$$

$$= \frac{\pi P}{2\eta l}\left(\frac{R^2 r^2}{2} - \frac{r^4}{4}\right)\Big|_0^R = \frac{\pi P R^4}{8\eta l}.$$

(b) Since

$$\text{Rate of blood flow} = \frac{\pi P R^4}{8\eta l},$$

if we take $k = \pi P/(8\eta l)$, then we have

$$\text{Rate of blood flow} = kR^4,$$

that is, rate of blood flow is proportional to R^4, in accordance with Poiseuille's Law.

33. Any small piece of mass ΔM on either of the two spheres has kinetic energy $\frac{1}{2}v^2\Delta M$. Since the angular velocity of the two spheres is the same, the actual velocity of the piece ΔM will depend on how far away it is from the axis of revolution. The further away a piece is from the axis, the faster it must be moving and the larger its velocity v. This is because if ΔM is at a distance r from the axis, in one revolution it must trace out a circular path of length $2\pi r$ about the axis. Since every piece in either sphere takes 1 minute to make 1 revolution, pieces farther from the axis must move faster, as they travel a greater distance.

Thus, since the thin spherical shell has more of its mass concentrated farther from the axis of rotation than does the solid sphere, the bulk of it is traveling faster than the bulk of the solid sphere. So, it has the higher kinetic energy.

CAS Challenge Problems

37. **(a)** The expression for arc length in terms of a definite integral gives

$$A(t) = \int_0^t \sqrt{1 + \left(\frac{1}{2\sqrt{x}}\right)^2}\, dx = \frac{2\sqrt{t}\sqrt{1 + 4t} + \operatorname{arcsinh}(2\sqrt{t})}{4}.$$

The integral was evaluated using a computer algebra system; different systems may give the answer in different forms. Some may involve ln instead of arcsinh, which is the inverse function of the hyperbolic sine function.

(b) Figure 8.19 shows that the graphs of $A(t)$ and the graph of $y = t$ look very similar. This suggests that $A(t) \approx t$.

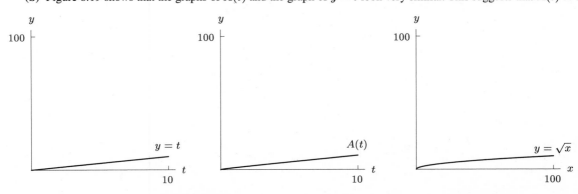

Figure 8.18   Figure 8.19

(c) The graph in Figure 8.19 is approximately horizontal and close to the x-axis. Thus, if we measure the arc length up to a certain x-value, the answer is approximately the same as if we had measured the length straight along the x-axis. Hence

$$A(t) \approx x = t.$$

So

$$A(t) \approx t.$$

CHECK YOUR UNDERSTANDING

1. True. Since $y = \pm\sqrt{9 - x^2}$ represent the top and bottom halves of the sphere, slicing disks perpendicular to the x-axis gives

$$\text{Volume of slice} \approx \pi y^2 \Delta x = \pi(9 - x^2)\Delta x$$
$$\text{Volume} = \int_{-3}^{3} \pi(9 - x^2)\, dx.$$

5. False. Volume is always positive, like area.

9. True. One way to look at it is that the center of mass shouldn't change if you change the units by which you measure the masses. If you double the masses, that is no different than using as a new unit of mass half the old unit. Alternatively, let the masses be m_1, m_2, and m_3 located at x_1, x_2, and x_3. Then the center of mass is given by:

$$\bar{x} = \frac{x_1 m_1 + x_2 m_2 + x_3 m_3}{m_1 + m_2 + m_3}.$$

Doubling the masses does not change the center of mass, since it doubles both the numerator and the denominator.

13. False. Work is the product of force and distance moved, so the work done in either case is 200 ft-lb.

17. False. The pressure is positive and when integrated gives a positive force.

21. False. It is true that $p(x) \geq 0$ for all x, but we also need $\int_{-\infty}^{\infty} p(x)dx = 1$. Since $p(x) = 0$ for $x \leq 0$, we need only check the integral from 0 to ∞. We have

$$\int_{0}^{\infty} xe^{-x^2}\, dx = \lim_{b \to \infty} \left(-\frac{1}{2}e^{-x^2}\right)\Big|_{0}^{b} = \frac{1}{2}.$$

25. False. Since f is concave down, this means that $f'(x)$ is decreasing, so $f'(x) \leq f'(0) = 3/4$ on the interval $[0, 4]$. However, it could be that $f'(x)$ becomes negative so that $(f'(x))^2$ becomes large, making the integral for the arc length large also. For example, $f(x) = (3/4)x - x^2$ is concave down and $f'(0) = 3/4$, but $f(0) = 0$ and $f(4) = -13$, so the graph of f on the interval $[0, 4]$ has arc length at least 13.

29. False. Note that p is the density function for the population, not the cumulative density function. Thus $p(10) = p(20)$ means that x values near 10 are as likely as x values near 20.

CHAPTER NINE

Solutions for Section 9.1

Exercises

1. Yes, $a = 1$, ratio $= -1/2$.

5. No. Ratio between successive terms is not constant: $\dfrac{2x^2}{x} = 2x$, while $\dfrac{3x^3}{2x^2} = \dfrac{3}{2}x$.

9. No. Ratio between successive terms is not constant: $\dfrac{6z^2}{3z} = 2z$, while $\dfrac{9z^3}{6z^2} = \dfrac{3}{2}z$.

13. Sum $= \dfrac{1}{1 - (-y^2)} = \dfrac{1}{1 + y^2}, |y| < 1.$

17. Using the formula for the sum of an infinite geometric series,

$$\sum_{n=4}^{\infty} \left(\frac{1}{3}\right)^n = \left(\frac{1}{3}\right)^4 + \left(\frac{1}{3}\right)^5 + \cdots = \left(\frac{1}{3}\right)^4 \left(1 + \frac{1}{3} + \left(\frac{1}{3}\right)^2 + \cdots\right) = \frac{\left(\frac{1}{3}\right)^4}{1 - \frac{1}{3}} = \frac{1}{54}$$

Problems

21. (a)

$$P_1 = 0$$
$$P_2 = 250(0.04)$$
$$P_3 = 250(0.04) + 250(0.04)^2$$
$$P_4 = 250(0.04) + 250(0.04)^2 + 250(0.04)^3$$
$$\vdots$$
$$P_n = 250(0.04) + 250(0.04)^2 + 250(0.04)^3 + \cdots + 250(0.04)^{n-1}$$

(b) $P_n = 250(0.04)\left(1 + (0.04) + (0.04)^2 + (0.04)^3 + \cdots + (0.04)^{n-2}\right) = 250\dfrac{0.04(1 - (0.04)^{n-1})}{1 - 0.04}$

(c)

$$P = \lim_{n \to \infty} P_n$$
$$= \lim_{n \to \infty} 250\frac{0.04(1 - (0.04)^{n-1})}{1 - 0.04}$$
$$= \frac{(250)(0.04)}{0.96} = 0.04Q \approx 10.42$$

Thus, $\lim_{n \to \infty} P_n = 10.42$ and $\lim_{n \to \infty} Q_n = 260.42$. We would expect these limits to differ because one is right before taking a tablet, one is right after. We would expect the difference between them to be 250 mg, the amount of ampicillin in one tablet.

25. (a)

$$\text{Total amount of money deposited} = 100 + 92 + 84.64 + \cdots$$
$$= 100 + 100(0.92) + 100(0.92)^2 + \cdots$$
$$= \frac{100}{1 - 0.92} = 1250 \quad \text{dollars}$$

(b) Credit multiplier $= 1250/100 = 12.50$

The 12.50 is the factor by which the bank has increased its deposits, from \$100 to \$1250.

Solutions for Section 9.2

Exercises

1. Since $\lim\limits_{n \to \infty} x^n = 0$ if $|x| < 1$ and $|0.2| < 1$, we have $\lim\limits_{n \to \infty} (0.2)^n = 0$.

5. Since $S_n = \cos(\pi n) = 1$ if n is even and $S_n = \cos(\pi n) = -1$ if n is odd, the values of S_n oscillate between 1 and -1, so the limit does not exist.

9. We use the integral test to determine whether this series converges or diverges. We determine whether the corresponding improper integral $\int_1^\infty \dfrac{1}{x^3}\,dx$ converges or diverges:

$$\int_1^\infty \frac{1}{x^3}\,dx = \lim_{b \to \infty} \int_1^b \frac{1}{x^3}\,dx = \lim_{b \to \infty} \left.\frac{-1}{2x^2}\right|_1^b = \lim_{b \to \infty} \left(\frac{-1}{2b^2} + \frac{1}{2}\right) = \frac{1}{2}.$$

Since the integral $\int_1^\infty \dfrac{1}{x^3}\,dx$ converges, we conclude from the integral test that the series $\sum\limits_{n=1}^{\infty} \dfrac{1}{n^3}$ converges.

Problems

13. The series $\sum\limits_{n=1}^{\infty} \left(\dfrac{3}{4}\right)^n$ is a convergent geometric series, but $\sum\limits_{n=1}^{\infty} \dfrac{1}{n}$ is the divergent harmonic series.

If $\sum\limits_{n=1}^{\infty} \left(\left(\dfrac{3}{4}\right)^n + \dfrac{1}{n}\right)$ converged, then $\sum\limits_{n=1}^{\infty} \left(\left(\dfrac{3}{4}\right)^n + \dfrac{1}{n}\right) - \sum\limits_{n=1}^{\infty} \left(\dfrac{3}{4}\right)^n = \sum\limits_{n=1}^{\infty} \dfrac{1}{n}$ would converge by Theorem 9.2.

Therefore $\sum\limits_{n=1}^{\infty} \left(\left(\dfrac{3}{4}\right)^n + \dfrac{1}{n}\right)$ diverges.

17. We use the integral test and calculate the corresponding improper integral, $\int_1^\infty 3/(2x - 1)^2\,dx$:

$$\int_1^\infty \frac{3\,dx}{(2x-1)^2} = \lim_{b \to \infty} \int_1^b \frac{3\,dx}{(2x-1)^2} = \lim_{b \to \infty} \left.\frac{-3/2}{(2x-1)}\right|_1^b = \lim_{b \to \infty} \left(\frac{-3/2}{(2b-1)} + \frac{3}{2}\right) = \frac{3}{2}.$$

Since the integral converges, the series $\sum\limits_{n=1}^{\infty} \dfrac{3}{(2n-1)^2}$ converges.

21. Using left-hand sums for the integral of $f(x) = 1/(4x - 3)$ over the interval $1 \le x \le n + 1$ with uniform subdivisions of length 1 gives a lower bound on the partial sum:

$$S_n = 1 + \frac{1}{5} + \frac{1}{9} + \cdots + \frac{1}{4n-3} > \int_1^{n+1} \frac{dx}{4x-3} = \left.\frac{1}{4}\ln(4x-3)\right|_1^{n+1} = \frac{1}{4}\ln(4n+1).$$

Since $\ln(4n + 1)$ increases without bound as $n \to \infty$, the partial sums of the series are unbounded. Thus, this is not a convergent series.

25. We want to define $\lim\limits_{n \to \infty} S_n = L$ so that S_n is as close to L as we please for all sufficiently large n. Thus, the definition says that for any positive ϵ, there is a value N such that

$$|S_n - L| < \epsilon \quad \text{whenever} \quad n \ge N.$$

29. From Property 1 in Theorem 9.2, we know that if $\sum a_n$ converges, then so does $\sum ka_n$.

Now suppose that $\sum a_n$ diverges and $\sum ka_n$ converges for $k \ne 0$. Thus using Property 1 and replacing $\sum a_n$ by $\sum ka_n$, we know that the following series converges:

$$\sum \frac{1}{k}(ka_n) = \sum a_n.$$

Thus, we have arrived at a contradiction, which means our original assumption, that $\sum\limits_{n=1}^{\infty} ka_n$ converged, must be wrong.

33. (a) Let N an integer with $N \geq c$. Consider the series $\sum_{i=N+1}^{\infty} a_i$. The partial sums of this series are increasing because all the terms in the series are positive. We show the partial sums are bounded using the right-hand sum in Figure 9.1. We see that for each positive integer k

$$f(N+1) + f(N+2) + \cdots + f(N+k) \leq \int_{N}^{N+k} f(x)\,dx.$$

Since $f(n) = a_n$ for all n, and $c \leq N$, we have

$$a_{N+1} + a_{N+2} + \cdots + a_{N+k} \leq \int_{c}^{N+k} f(x)\,dx.$$

Since $f(x)$ is a positive function, $\int_{c}^{N+k} f(x)\,dx \leq \int_{c}^{b} f(x)\,dx$ for all $b \geq N+k$. Since f is positive and $\int_{c}^{\infty} f(x)\,dx$ is convergent, $\int_{c}^{N+k} f(x)\,dx < \int_{c}^{\infty} f(x)\,dx$, so we have

$$a_{N+1} + a_{N+2} + \cdots + a_{N+k} \leq \int_{c}^{\infty} f(x)\,dx \quad \text{for all } k.$$

Thus, the partial sums of the series $\sum_{i=N+1}^{\infty} a_i$ are all bounded by the same number, so this series converges. Now use Theorem 9.2, property 2, to conclude that $\sum_{i=1}^{\infty} a_i$ converges.

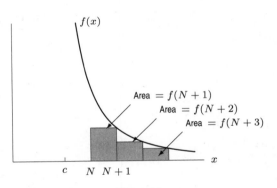

Figure 9.1

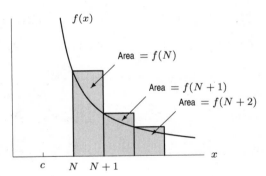

Figure 9.2

(b) We now suppose $\int_{c}^{\infty} f(x)\,dx$ diverges. In Figure 9.2 we see that for each positive integer k

$$\int_{N}^{N+k+1} f(x)\,dx \leq f(N) + f(N+1) + \cdots + f(N+k).$$

Since $f(n) = a_n$ for all n, we have

$$\int_{N}^{N+k+1} f(x)\,dx \leq a_N + a_{N+1} + \cdots + a_{N+k}.$$

Since $f(x)$ is defined for all $x \geq c$, if $\int_{c}^{\infty} f(x)\,dx$ is divergent, then $\int_{N}^{\infty} f(x)\,dx$ is divergent. So as $k \to \infty$, the the integral $\int_{N}^{N+k+1} f(x)\,dx$ diverges, so the partial sums of the series $\sum_{i=N}^{\infty} a_i$ diverge. Thus, the series $\sum_{i=1}^{\infty} a_i$ diverges.

More precisely, suppose the series converged. Then the partial sums would be bounded. (The partial sums would be less than the sum of the series, since all the terms in the series are positive.) But that would imply that the integral converged, by Theorem 9.1 on Convergence of Increasing Bounded Sequences. This contradicts the assumption that $\int_{N}^{\infty} f(x)\,dx$ is divergent.

Solutions for Section 9.3

Exercises

1. Let $a_n = 1/(n^2 + 2)$. Since $n^2 + 2 > n^2$, we have $1/(n^2 + 2) < 1/n^2$, so

$$0 < a_n < \frac{1}{n^2}.$$

The series $\displaystyle\sum_{n=1}^{\infty} \frac{1}{n^2}$ converges, so the comparison test tells us that the series $\displaystyle\sum_{n=1}^{\infty} \frac{1}{n^2 + 2}$ also converges.

5. Since $\ln n \leq n$ for $n \geq 2$, we have $1/\ln n \geq 1/n$, so the series diverges by comparison with the harmonic series, $\sum 1/n$.

9. Let $a_n = n^2/(n^4 + 1)$. Since $n^4 + 1 > n^4$, we have $\dfrac{1}{n^4 + 1} < \dfrac{1}{n^4}$, so

$$a_n = \frac{n^2}{n^4 + 1} < \frac{n^2}{n^4} = \frac{1}{n^2},$$

therefore

$$0 < a_n < \frac{1}{n^2}.$$

Since the series $\displaystyle\sum_{n=1}^{\infty} \frac{1}{n^2}$ converges, the comparison test tells us that the series $\displaystyle\sum_{n=1}^{\infty} \frac{n^2}{n^4 + 1}$ converges also.

13. Since $a_n = (n!)^2/(2n)!$, replacing n by $n + 1$ gives $a_{n+1} = ((n+1)!)^2/(2n+2)!$. Thus,

$$\frac{|a_{n+1}|}{|a_n|} = \frac{\dfrac{((n+1)!)^2}{(2n+2)!}}{\dfrac{(n!)^2}{(2n)!}} = \frac{((n+1)!)^2}{(2n+2)!} \cdot \frac{(2n)!}{(n!)^2}.$$

However, since $(n+1)! = (n+1)n!$ and $(2n+2)! = (2n+2)(2n+1)(2n)!$, we have

$$\frac{|a_{n+1}|}{|a_n|} = \frac{(n+1)^2(n!)^2(2n)!}{(2n+2)(2n+1)(2n)!(n!)^2} = \frac{(n+1)^2}{(2n+2)(2n+1)} = \frac{n+1}{4n+2},$$

so

$$L = \lim_{n \to \infty} \frac{|a_{n+1}|}{|a_n|} = \frac{1}{4}.$$

Since $L < 1$, the ratio test tells us that $\displaystyle\sum_{n=1}^{\infty} \frac{(n!)^2}{(2n)!}$ converges.

17. Let $a_n = 1/\sqrt{n}$. Then replacing n by $n+1$ we have $a_{n+1} = 1/\sqrt{n+1}$. Since $\sqrt{n+1} > \sqrt{n}$, we have $\dfrac{1}{\sqrt{n+1}} < \dfrac{1}{\sqrt{n}}$, hence $a_{n+1} < a_n$. In addition, $\lim_{n \to \infty} a_n = 0$ so $\displaystyle\sum_{n=0}^{\infty} \frac{(-1)^n}{\sqrt{n}}$ converges by the alternating series test.

Problems

21. The first few terms of the series may be written

$$1 + e^{-1} + e^{-2} + e^{-3} + \cdots;$$

this is a geometric series with $a = 1$ and $x = e^{-1} = 1/e$. Since $|x| < 1$, the geometric series converges to

$$S = \frac{1}{1 - x} = \frac{1}{1 - e^{-1}} = \frac{e}{e - 1}.$$

25. Let $a_n = 1/\sqrt{3n - 1}$. Then replacing n by $n + 1$ gives $a_{n+1} = 1/\sqrt{3(n + 1) - 1}$. Since

$$\sqrt{3(n + 1) - 1} > \sqrt{3n - 1},$$

we have

$$a_{n+1} < a_n.$$

In addition, $\lim_{n \to \infty} a_n = 0$ so the alternating series test tells us that the series $\sum_{n=1}^{\infty} \dfrac{(-1)^{n-1}}{\sqrt{3n - 1}}$ converges.

29. **(a)** Assume that n is even. Then

$$1 - \frac{1}{2} + \frac{1}{3} - \frac{1}{4} + \cdots - \frac{1}{n} = \left(1 - \frac{1}{2}\right) + \left(\frac{1}{3} - \frac{1}{4}\right) + \cdots + \left(\frac{1}{n - 1} - \frac{1}{n}\right)$$

$$= \frac{1}{1 \cdot 2} + \frac{1}{3 \cdot 4} + \cdots + \frac{1}{(n - 1) \cdot n}.$$

(b) The given series $\dfrac{1}{1 \cdot 2} + \dfrac{1}{3 \cdot 4} + \dfrac{1}{5 \cdot 6} + \cdots$ is term by term less than the series

$$\frac{1}{1 \cdot 1} + \frac{1}{2 \cdot 2} + \frac{1}{3 \cdot 3} + \cdots = \frac{1}{1^2} + \frac{1}{2^2} + \frac{1}{3^3} + \cdots.$$

Since this second series, $\sum 1/n^2$, converges by the integral test, the first series converges.

(c) By parts (a) and (b), the sequence of partial sums for even n converges. The partial sum for odd n equals $1/n$ plus the partial sum for even $n - 1$. Thus the partial sums for odd n approach the partial sums for even n, as $n \to \infty$. Therefore the sequence of all partial sums converges, and hence the series converges.

Solutions for Section 9.4

Exercises

1. Yes.

5. The general term can be written as $\dfrac{1 \cdot 3 \cdot 5 \cdots (2n - 1)}{2^n \cdot n!} x^n$ for $n \geq 1$.

9. The general term can be written as $\dfrac{(x - a)^n}{2^{n-1} \cdot n!}$ for $n \geq 1$.

13. Since $C_n = (n + 1)/(2^n + n)$, replacing n by $n + 1$ gives $C_{n+1} = (n + 2)/(2^{n+1} + n + 1)$. Using the ratio test, we have

$$\frac{|a_{n+1}|}{|a_n|} = |x|\frac{|C_{n+1}|}{|C_n|} = |x|\frac{(n + 2)/(2^{n+1} + n + 1)}{(n + 1)/(2^n + n)} = |x|\frac{n + 2}{2^{n+1} + n + 1} \cdot \frac{2^n + n}{n + 1} = |x|\frac{n + 2}{n + 1} \cdot \frac{2^n + n}{2^{n+1} + n + 1}.$$

Since

$$\lim_{n \to \infty} \frac{n + 2}{n + 1} = 1$$

and

$$\lim_{n \to \infty} \left(\frac{2^n + n}{2^{n+1} + n + 1}\right) = \frac{1}{2} \lim_{n \to \infty} \left(\frac{2^n + n}{2^n + (n + 1)/2}\right) = \frac{1}{2},$$

because 2^n dominates n as $n \to \infty$, we have

$$\lim_{n \to \infty} \frac{|a_{n+1}|}{|a_n|} = \frac{1}{2}|x|.$$

Thus the radius of convergence is $R = 2$.

17. Here the coefficient of the nth term is $C_n = (2^n/n!)$. Now we have

$$\left|\frac{a_{n+1}}{a_n}\right| = \left|\frac{(2^{n+1}/(n + 1)!)x^{n+1}}{(2^n/n!)x^n}\right| = \frac{2|x|}{n + 1} \to 0 \text{ as } n \to \infty.$$

Thus, the radius of convergence is $R = \infty$, and the series converges for all x.

21. We write the series as

$$x - \frac{x^3}{3} + \frac{x^5}{5} - \frac{x^7}{7} + \cdots + (-1)^{n-1}\frac{x^{2n-1}}{2n-1} + \cdots,$$

so

$$a_n = (-1)^{n-1}\frac{x^{2n-1}}{2n-1}.$$

Replacing n by $n+1$, we have

$$a_{n+1} = (-1)^{n+1-1}\frac{x^{2(n+1)-1}}{2(n+1)-1} = (-1)^n\frac{x^{2n+1}}{2n+1}.$$

Thus

$$\frac{|a_{n+1}|}{|a_n|} = \left|\frac{(-1)^n x^{2n+1}}{2n+1}\right| \cdot \left|\frac{2n-1}{(-1)^{n-1}x^{2n-1}}\right| = \frac{2n-1}{2n+1}x^2,$$

so

$$L = \lim_{n\to\infty}\frac{|a_{n+1}|}{|a_n|} = \lim_{n\to\infty}\frac{2n-1}{2n+1}x^2 = x^2.$$

By the ratio test, this series converges if $L < 1$, that is, if $x^2 < 1$, so $R = 1$.

Problems

25. The kth coefficient in the series $\sum kC_k x^k$ is $D_k = k \cdot C_k$. We are given that the series $\sum C_k x^k$ has radius of convergence R by the ratio test, so

$$|x|\lim_{k\to\infty}\frac{|C_{k+1}|}{|C_k|} = \frac{|x|}{R}.$$

Thus, applying the ratio test to the new series, we have

$$\lim_{k\to\infty}\left|\frac{D_{k+1}x^{k+1}}{D_k x^k}\right| = \lim_{k\to\infty}\left|\frac{(k+1)C_{k+1}}{kC_k}\right||x| = \frac{|x|}{R}.$$

Hence the new series has radius of convergence R.

Solutions for Chapter 9 Review

Exercises

1. Let $a_n = n^2/(3n^2 + 4)$. Since $3n^2 + 4 > 3n^2$, we have $\dfrac{n^2}{3n^2+4} < \dfrac{1}{3}$, so

$$0 < a_n < \left(\frac{1}{3}\right)^n.$$

The geometric series $\displaystyle\sum_{n=1}^{\infty}\left(\frac{1}{3}\right)^n$ converges, so the comparison test tells us that the series $\displaystyle\sum_{n=1}^{\infty}\left(\frac{n^2}{3n^2+4}\right)^n$ also converges.

5. To show that the original series converges, we show that the series $\displaystyle\sum_{n=1}^{\infty}\left(\frac{3}{4}\right)^n$ and $\displaystyle\sum_{n=1}^{\infty}\frac{1}{n^2}$ converge. The first of these is a convergent geometric series, since $|3/4| < 1$. The integral test tells us that series $\displaystyle\sum_{n=1}^{\infty}\frac{1}{n^2}$ converges by comparing it with the convergent integral $\int_0^1 1/x^2\,dx$. Theorem 9.2 then tells us that the series $\displaystyle\sum_{n=1}^{\infty}\left(\frac{3}{4}^n + \frac{1}{n^2}\right)$ also converges.

9. Since $a_n = (2n)!/(n!)^2$, replacing n by $n+1$ gives $a_{n+1} = (2n+2)!/((n+1)!)^2$. Thus

$$\frac{a_{n+1}}{a_n} = \frac{\frac{(2n+2)!}{((n+1)!)^2}}{\frac{(2n)!}{(n!)^2}} = \frac{(2n+2)!}{(n+1)!(n+1)!} \cdot \frac{n!n!}{(2n)!}.$$

Since $(2n+2)! = (2n+2)(2n+1)(2n)!$ and $(n+1)! = (n+1)n!$, we have

$$\frac{a_{n+1}}{a_n} = \frac{(2n+2)(2n+1)}{(n+1)(n+1)},$$

therefore

$$L = \lim_{n \to \infty} \frac{a_{n+1}}{a_n} = 4.$$

As $L = 4$ the ratio test tells us that the series $\sum_{n=1}^{\infty} \frac{(2n)!}{(n!)^2}$ diverges.

13. Let $C_n = \frac{(2n)!}{(n!)^2}$. Then replacing n by $n+1$, we have $C_{n+1} = \frac{(2n+2)!}{((n+1)!)^2}$. Thus, with $a_n = (2n)!x^n/(n!)^2$, we have

$$\frac{|a_{n+1}|}{|a_n|} = |x|\frac{|C_{n+1}|}{|C_n|} = |x|\frac{(2n+2)!/((n+1)!)^2}{(2n)!/(n!)^2} = |x|\frac{(2n+2)!}{(2n)!} \cdot \frac{(n!)^2}{((n+1)!)^2}.$$

Since $(2n+2)! = (2n+2)(2n+1)(2n)!$ and $(n+1)! = (n+1)n!$ we have

$$\frac{|C_{n+1}|}{|C_n|} = \frac{(2n+2)(2n+1)}{(n+1)(n+1)},$$

so

$$\lim_{n \to \infty} \frac{|a_{n+1}|}{|a_n|} = |x| \lim_{n \to \infty} \frac{|C_{n+1}|}{|C_n|} = |x| \lim_{n \to \infty} \frac{(2n+2)(2n+1)}{(n+1)(n+1)} = |x| \lim_{n \to \infty} \frac{4n+2}{n+1} = 4|x|,$$

so the radius of convergence of this series is $R = 1/4$.

Problems

17. The amount of cephalexin in the body is given by $Q(t) = Q_0e^{-kt}$, where $Q_0 = Q(0)$ and k is a constant. Since the half-life is 0.9 hours,

$$\frac{1}{2} = e^{-0.9k}, \quad k = -\frac{1}{0.9} \ln \frac{1}{2} \approx 0.8.$$

(a) After 6 hours

$$Q = Q_0e^{-k(6)} \approx Q_0e^{-0.8(6)} = Q_0(0.01).$$

Thus, the percentage of the cephalexin that remains after 6 hours $\approx 1\%$.

(b)

$$Q_1 = 250$$
$$Q_2 = 250 + 250(0.01)$$
$$Q_3 = 250 + 250(0.01) + 250(0.01)^2$$
$$Q_4 = 250 + 250(0.01) + 250(0.01)^2 + 250(0.01)^3$$

(c)

$$Q_3 = \frac{250(1 - (0.01)^3)}{1 - 0.01}$$
$$\approx 252.5$$
$$Q_4 = \frac{250(1 - (0.01)^4)}{1 - 0.01}$$
$$\approx 252.5$$

Thus, by the time a patient has taken three cephalexin tablets, the quantity of drug in the body has leveled off to 252.5 mg.

(d) Looking at the answers to part (b) shows that

$$Q_n = 250 + 250(0.01) + 250(0.01)^2 + \cdots + 250(0.01)^{n-1}$$
$$= \frac{250(1 - (0.01)^n)}{1 - 0.01}.$$

(e) In the long run, $n \to \infty$. So,

$$Q = \lim_{n \to \infty} Q_n = \frac{250}{1 - 0.01} = 252.5.$$

21.

$$\text{Present value of first coupon} = \frac{50}{1.06}$$
$$\text{Present value of second coupon} = \frac{50}{(1.06)^2}, \text{ etc.}$$

$$\text{Total present value} = \underbrace{\frac{50}{1.06} + \frac{50}{(1.06)^2} + \cdots + \frac{50}{(1.06)^{10}}}_{\text{coupons}} + \underbrace{\frac{1000}{(1.06)^{10}}}_{\text{principal}}$$

$$= \frac{50}{1.06}\left(1 + \frac{1}{1.06} + \cdots + \frac{1}{(1.06)^9}\right) + \frac{1000}{(1.06)^{10}}$$

$$= \frac{50}{1.06}\left(\frac{1 - \left(\frac{1}{1.06}\right)^{10}}{1 - \frac{1}{1.06}}\right) + \frac{1000}{(1.06)^{10}}$$

$$= 368.004 + 558.395$$

$$= \$926.40$$

25. (a) Since

$$|a_n| = a_n \qquad \text{if } a_n \geq 0$$
$$|a_n| = -a_n \qquad \text{if } a_n < 0,$$

we have

$$a_n + |a_n| = 2|a_n| \qquad \text{if } a_n > 0$$
$$a_n + |a_n| = 0 \qquad \text{if } a_n < 0.$$

Thus, for all n,

$$0 \leq a_n + |a_n| \leq 2|a_n|.$$

(b) If $\sum |a_n|$ converges, then $\sum 2|a_n|$ is convergent, so, by comparison, $\sum (a_n + |a_n|)$ is convergent. Then

$$\sum ((a_n + |a_n|) - |a_n|) = \sum a_n$$

is convergent, as it is the difference of two convergent series.

CHECK YOUR UNDERSTANDING

1. True. A geometric series, $a + ax + ax^2 + \cdots$, is a power series about $x = 0$ with all coefficients equal to a.

5. True. This power series has an interval of convergence centered on $x = 0$. If the power series does not converge for $x = 1$, then the radius of convergence is less than or equal to 1. Thus, $x = 2$ lies outside the interval of convergence, so the series does not converge there.

9. False. It is true that if $\sum |a_n|$ converges, then we know that $\sum a_n$ converges. However, knowing that $\sum a_n$ converges does *not* tell us that $\sum |a_n|$ converges.

For example, if $a_n = (-1)^{n-1}/n$, then $\sum a_n$ converges by the alternating series test. However, $\sum |a_n|$ is the harmonic series which diverges.

13. True. Writing out the terms of this series, we have

$$(1 + (-1)^1) + (1 + (-1)^2) + (1 + (-1)^3) + (1 + (-1)^4) + \cdots$$
$$= (1 - 1) + (1 + 1) + (1 - 1) + (1 + 1) + \cdots$$
$$= 0 + 2 + 0 + 2 + \cdots.$$

17. True. Let $c_n = (-1)^n |a_n|$. Then $|c_n| = |a_n|$ so $\sum |c_n|$ converges, and therefore $\sum c_n = \sum (-1)^n |a_n|$ converges.

21. False. Consider the power series

$$(x - 1) - \frac{(x - 1)^2}{2} + \frac{(x - 1)^3}{3} + \cdots + (-1)^{n-1} \frac{(x - 1)^n}{n} + \cdots,$$

whose interval of convergence is $0 < x \leq 2$. This series converges at one endpoint, $x = 2$, but not at the other, $x = 0$.

25. False. If $a_n b_n = 1/n^2$ and $a_n = b_n = 1/n$, then $\sum a_n b_n$ converges, but $\sum a_n$ and $\sum b_n$ do not converge.

CHAPTER TEN

Solutions for Section 10.1

Exercises

1. Let $\dfrac{1}{1+x} = (1+x)^{-1}$. Then $f(0) = 1$.

$$
\begin{array}{ll}
f'(x) = -1!(1+x)^{-2} & f'(0) = -1, \\
f''(x) = 2!(1+x)^{-3} & f''(0) = 2!, \\
f'''(x) = -3!(1+x)^{-4} & f'''(0) = -3!, \\
f^{(4)}(x) = 4!(1+x)^{-5} & f^{(4)}(0) = 4!, \\
f^{(5)}(x) = -5!(1+x)^{-6} & f^{(5)}(0) = -5!, \\
f^{(6)}(x) = 6!(1+x)^{-7} & f^{(6)}(0) = 6!, \\
f^{(7)}(x) = -7!(1+x)^{-8} & f^{(7)}(0) = -7!, \\
f^{(8)}(x) = 8!(1+x)^{-9} & f^{(8)}(0) = 8!.
\end{array}
$$

$$
\begin{aligned}
P_4(x) &= 1 - x + x^2 - x^3 + x^4, \\
P_6(x) &= 1 - x + x^2 - x^3 + x^4 - x^5 + x^6, \\
P_8(x) &= 1 - x + x^2 - x^3 + x^4 - x^5 + x^6 - x^7 + x^8.
\end{aligned}
$$

5. Let $f(x) = \arctan x$. Then $f(0) = \arctan 0 = 0$, and

$$
\begin{array}{ll}
f'(x) = 1/(1+x^2) = (1+x^2)^{-1} & f'(0) = 1, \\
f''(x) = (-1)(1+x^2)^{-2}2x & f''(0) = 0, \\
f'''(x) = 2!(1+x^2)^{-3}2^2x^2 + (-1)(1+x^2)^{-2}2 & f'''(0) = -2, \\
f^{(4)}(x) = -3!(1+x^2)^{-4}2^3x^3 + 2!(1+x^2)^{-3}2^3x & \\
\qquad + 2!(1+x^2)^{-3}2^2x & f^{(4)}(0) = 0.
\end{array}
$$

Therefore,

$$
P_3(x) = P_4(x) = x - \frac{1}{3}x^3.
$$

9. Let $f(x) = \dfrac{1}{\sqrt{1+x}} = (1+x)^{-1/2}$. Then $f(0) = 1$.

$$
\begin{array}{ll}
f'(x) = -\frac{1}{2}(1+x)^{-3/2} & f'(0) = -\frac{1}{2}, \\
f''(x) = \frac{3}{2^2}(1+x)^{-5/2} & f''(0) = \frac{3}{2^2}, \\
f'''(x) = -\frac{3 \cdot 5}{2^3}(1+x)^{-7/2} & f'''(0) = -\frac{3 \cdot 5}{2^3}, \\
f^{(4)}(x) = \frac{3 \cdot 5 \cdot 7}{2^4}(1+x)^{-9/2} & f^{(4)}(0) = \frac{3 \cdot 5 \cdot 7}{2^4}
\end{array}
$$

Then,

$$
\begin{aligned}
P_2(x) &= 1 - \frac{1}{2}x + \frac{1}{2!}\frac{3}{2^2}x^2 = 1 - \frac{1}{2}x + \frac{3}{8}x^2, \\
P_3(x) &= P_2(x) - \frac{1}{3!}\frac{3 \cdot 5}{2^3}x^3 = 1 - \frac{1}{2}x + \frac{3}{8}x^2 - \frac{5}{16}x^3, \\
P_4(x) &= P_3(x) + \frac{1}{4!}\frac{3 \cdot 5 \cdot 7}{2^4}x^4 = 1 - \frac{1}{2}x + \frac{3}{8}x^2 - \frac{5}{16}x^3 + \frac{35}{128}x^4.
\end{aligned}
$$

13. Let $f(x) = e^x$. Since $f^{(k)}(x) = e^x = f(x)$ for all $k \geq 1$, the Taylor polynomial of degree 4 for $f(x) = e^x$ about $x = 1$ is

$$
\begin{aligned}
P_4(x) &= e^1 + e^1(x-1) + \frac{e^1}{2!}(x-1)^2 + \frac{e^1}{3!}(x-1)^3 + \frac{e^1}{4!}(x-1)^4 \\
&= e\left[1 + (x-1) + \frac{1}{2}(x-1)^2 + \frac{1}{6}(x-1)^3 + \frac{1}{24}(x-1)^4\right].
\end{aligned}
$$

Problems

17. As we can see from Problem 15, a is the y-intercept of $f(x)$, b is the slope of the tangent line to $f(x)$ at $x = 0$ and c tells us the concavity of $f(x)$ near $x = 0$.

So $a < 0$, $b > 0$ and $c > 0$.

21.

$$f(x) = 4x^2 - 7x + 2 \quad f(0) = 2$$
$$f'(x) = 8x - 7 \qquad f'(0) = -7$$
$$f''(x) = 8 \qquad f''(0) = 8,$$

so $P_2(x) = 2 + (-7)x + \frac{8}{2}x^2 = 4x^2 - 7x + 2$. We notice that $f(x) = P_2(x)$ in this case.

25.

$$\lim_{x \to 0} \frac{1 - \cos x}{x^2} = \lim_{x \to 0} \frac{1 - (1 - \frac{x^2}{2!} + \frac{x^4}{4!})}{x^2} = \lim_{x \to 0} \left(\frac{1}{2} - \frac{x^2}{4!} \right) = \frac{1}{2}.$$

29. (a) $f(x) = e^{x^2}$.

$f'(x) = 2xe^{x^2}$, $f''(x) = 2(1 + 2x^2)e^{x^2}$, $f'''(x) = 4(3x + 2x^3)e^{x^2}$,
$f^{(4)}(x) = 4(3 + 6x^2)e^{x^2} + 4(3x + 2x^3)2xe^{x^2}$.
The Taylor polynomial about $x = 0$ is

$$P_4(x) = 1 + \frac{0}{1!}x + \frac{2}{2!}x^2 + \frac{0}{3!}x^3 + \frac{12}{4!}x^4$$
$$= 1 + x^2 + \frac{1}{2}x^4.$$

(b) $f(x) = e^x$. The Taylor polynomial of degree 2 is

$$Q_2(x) = 1 + \frac{x}{1!} + \frac{x^2}{2!} = 1 + x + \frac{1}{2}x^2.$$

If we substitute x^2 for x in the Taylor polynomial for e^x of degree 2, we will get $P_4(x)$, the Taylor polynomial for e^{x^2} of degree 4:

$$Q_2(x^2) = 1 + x^2 + \frac{1}{2}(x^2)^2$$
$$= 1 + x^2 + \frac{1}{2}x^4$$
$$= P_4(x).$$

(c) Let $Q_{10}(x) = 1 + \frac{x}{1!} + \frac{x^2}{2!} + \cdots + \frac{x^{10}}{10!}$ be the Taylor polynomial of degree 10 for e^x about $x = 0$. Then

$$P_{20}(x) = Q_{10}(x^2)$$
$$= 1 + \frac{x^2}{1!} + \frac{(x^2)^2}{2!} + \cdots + \frac{(x^2)^{10}}{10!}$$
$$= 1 + \frac{x^2}{1!} + \frac{x^4}{2!} + \cdots + \frac{x^{20}}{10!}.$$

(d) Let $e^x \approx Q_5(x) = 1 + \frac{x}{1!} + \cdots + \frac{x^5}{5!}$. Then

$$e^{-2x} \approx Q_5(-2x)$$
$$= 1 + \frac{-2x}{1!} + \frac{(-2x)^2}{2!} + \frac{(-2x)^3}{3!} + \frac{(-2x)^4}{4!} + \frac{(-2x)^5}{5!}$$
$$= 1 - 2x + 2x^2 - \frac{4}{3}x^3 + \frac{2}{3}x^4 - \frac{4}{15}x^5.$$

Solutions for Section 10.2

Exercises

1.

$$f(x) = \tfrac{1}{1-x} = (1-x)^{-1} \qquad\qquad f(0) = 1,$$
$$f'(x) = -(1-x)^{-2}(-1) = (1-x)^{-2} \qquad f'(0) = 1,$$
$$f''(x) = -2(1-x)^{-3}(-1) = 2(1-x)^{-3} \qquad f''(0) = 2,$$
$$f'''(x) = -6(1-x)^{-4}(-1) = 6(1-x)^{-4} \qquad f'''(0) = 6.$$

$$f(x) = \frac{1}{1-x} = 1 + 1 \cdot x + \frac{2x^2}{2!} + \frac{6x^3}{3!} + \cdots$$
$$= 1 + x + x^2 + x^3 + \cdots$$

5.

$$f(x) = \sin x \qquad\qquad f(\tfrac{\pi}{4}) = \tfrac{\sqrt{2}}{2},$$
$$f'(x) = \cos x \qquad\qquad f'(\tfrac{\pi}{4}) = \tfrac{\sqrt{2}}{2},$$
$$f''(x) = -\sin x \qquad\qquad f''(\tfrac{\pi}{4}) = -\tfrac{\sqrt{2}}{2},$$
$$f'''(x) = -\cos x \qquad\qquad f'''(\tfrac{\pi}{4}) = -\tfrac{\sqrt{2}}{2}.$$

$$\sin x = \frac{\sqrt{2}}{2} + \frac{\sqrt{2}}{2}\left(x - \frac{\pi}{4}\right) - \frac{\sqrt{2}}{2}\frac{(x - \frac{\pi}{4})^2}{2!} - \frac{\sqrt{2}}{2}\frac{(x - \frac{\pi}{4})^3}{3!} - \cdots$$
$$= \frac{\sqrt{2}}{2} + \frac{\sqrt{2}}{2}\left(x - \frac{\pi}{4}\right) - \frac{\sqrt{2}}{4}\left(x - \frac{\pi}{4}\right)^2 - \frac{\sqrt{2}}{12}\left(x - \frac{\pi}{4}\right)^3 - \cdots$$

9.

$$f(x) = \tfrac{1}{x} \qquad\qquad f(1) = 1$$
$$f'(x) = -\tfrac{1}{x^2} \qquad f'(1) = -1$$
$$f''(x) = \tfrac{2}{x^3} \qquad f''(1) = 2$$
$$f'''(x) = -\tfrac{6}{x^4} \qquad f'''(1) = -6$$

$$\frac{1}{x} = 1 - (x-1) + \frac{2(x-1)^2}{2!} - \frac{6(x-1)^3}{3!} + \cdots$$
$$= 1 - (x-1) + (x-1)^2 - (x-1)^3 + \cdots.$$

13. The general term can be written as $(-1)^n x^n$ for $n \geq 0$.

17. The general term can be written as $(-1)^k x^{2k+1}/(2k+1)$ for $k \geq 0$.

Problems

21. (a)
$$f(x) = \ln(1 + 2x) \qquad f(0) = 0$$
$$f'(x) = \tfrac{2}{1+2x} \qquad f'(0) = 2$$
$$f''(x) = -\tfrac{4}{(1+2x)^2} \qquad f''(0) = -4$$
$$f'''(x) = \tfrac{16}{(1+2x)^3} \qquad f'''(0) = 16$$

$$\ln(1 + 2x) = 2x - 2x^2 + \frac{8}{3}x^3 + \cdots$$

(b) To get the expression for $\ln(1 + 2x)$ from the series for $\ln(1 + x)$, substitute $2x$ for x in the series

$$\ln(1 + x) = x - \frac{x^2}{2} + \frac{x^3}{3} - \frac{x^4}{4} + \cdots$$

to get

$$\ln(1 + 2x) = 2x - \frac{(2x)^2}{2} + \frac{(2x)^3}{3} - \frac{(2x)^4}{4} + \cdots$$

$$= 2x - 2x^2 + \frac{8x^3}{3} - 4x^4 + \cdots$$

(c) Since the interval of convergence for $\ln(1 + x)$ is $-1 < x < 1$, substituting $2x$ for x suggests the interval of convergence of $\ln(1 + 2x)$ is $-1 < 2x < 1$, or $-\frac{1}{2} < x < \frac{1}{2}$.

25. The Taylor series for $\ln(1 - x)$ is

$$\ln(1 - x) = -x - \frac{x^2}{2} - \frac{x^3}{3} - \cdots - \frac{x^n}{n} - \cdots,$$

so

$$\lim_{n \to \infty} \frac{|a_{n+1}|}{|a_n|} = |x| \lim_{n \to \infty} \frac{1/(n+1)}{1/n} = |x| \lim_{n \to \infty} \left| \frac{n}{n+1} \right| = |x|.$$

Thus the series converges for $|x| < 1$, and the radius of convergence is 1. Note: This series can be obtained from the series for $\ln(1 + x)$ by replacing x by $-x$ and has the same radius of convergence as the series for $\ln(1 + x)$.

29. This is the series for $1/(1 - x)$ with x replaced by $1/4$, so the series converges to $1/(1 - (1/4)) = 4/3$.

33. This is the series for e^x with $x = 3$ substituted. Thus

$$1 + 3 + \frac{9}{2!} + \frac{27}{3!} + \frac{81}{4!} + \cdots = 1 + 3 + \frac{3^2}{2!} + \frac{3^3}{3!} + \frac{3^4}{4!} + \cdots = e^3.$$

37. Since $x - \frac{1}{2}x^2 + \frac{1}{3}x^3 + \cdots = \ln(1 + x)$, we solve $\ln(1 + x) = 0.2$, giving $1 + x = e^{0.2}$, so $x = e^{0.2} - 1$.

41. We define $e^{i\theta}$ to be

$$e^{i\theta} = 1 + i\theta + \frac{(i\theta)^2}{2!} + \frac{(i\theta)^3}{3!} + \frac{(i\theta)^4}{4!} + \frac{(i\theta)^5}{5!} + \frac{(i\theta)^6}{6!} + \cdots$$

Suppose we consider the expression $\cos \theta + i \sin \theta$, with $\cos \theta$ and $\sin \theta$ replaced by their Taylor series:

$$\cos \theta + i \sin \theta = \left(1 - \frac{\theta^2}{2!} + \frac{\theta^4}{4!} - \frac{\theta^6}{6!} + \cdots \right) + i \left(\theta - \frac{\theta^3}{3!} + \frac{\theta^5}{5!} - \cdots \right)$$

Reordering terms, we have

$$\cos \theta + i \sin \theta = 1 + i\theta - \frac{\theta^2}{2!} - \frac{i\theta^3}{3!} + \frac{\theta^4}{4!} + \frac{i\theta^5}{5!} - \frac{\theta^6}{6!} - \cdots$$

Using the fact that $i^2 = -1$, $i^3 = -i$, $i^4 = 1$, $i^5 = i$, $\cdots$, we can rewrite the series as

$$\cos \theta + i \sin \theta = 1 + i\theta + \frac{(i\theta)^2}{2!} + \frac{(i\theta)^3}{3!} + \frac{(i\theta)^4}{4!} + \frac{(i\theta)^5}{5!} + \frac{(i\theta)^6}{6!} + \cdots$$

Amazingly enough, this series is the Taylor series for e^x with $i\theta$ substituted for x. Therefore, we have shown that

$$\cos \theta + i \sin \theta = e^{i\theta}.$$

Solutions for Section 10.3

Exercises

1. We'll use

$$\sqrt{1+y} = (1+y)^{\frac{1}{2}} = 1 + \left(\frac{1}{2}\right)y + \left(\frac{1}{2}\right)\left(\frac{-1}{2}\right)\frac{y^2}{2!}$$
$$+ \left(\frac{1}{2}\right)\left(\frac{-1}{2}\right)\left(\frac{-3}{2}\right)\frac{y^3}{3!} + \cdots$$
$$= 1 + \frac{y}{2} - \frac{y^2}{8} + \frac{y^3}{16} - \cdots.$$

Substitute $y = -2x$.

$$\sqrt{1-2x} = 1 + \frac{(-2x)}{2} - \frac{(-2x)^2}{8} + \frac{(-2x)^3}{16} - \cdots$$
$$= 1 - x - \frac{x^2}{2} - \frac{x^3}{2} - \cdots$$

5. Substituting $x = -2y$ into $\ln(1+x) = x - \frac{x^2}{2} + \frac{x^3}{3} - \frac{x^4}{4} + \cdots$ gives

$$\ln(1-2y) = (-2y) - \frac{(-2y)^2}{2} + \frac{(-2y)^3}{3} - \frac{(-2y)^4}{4} + \cdots$$
$$= -2y - 2y^2 - \frac{8}{3}y^3 - 4y^4 - \cdots.$$

9.

$$\frac{z}{e^{z^2}} = ze^{-z^2} = z\left(1 + (-z^2) + \frac{(-z^2)^2}{2!} + \frac{(-z^2)^3}{3!} + \cdots\right)$$
$$= z - z^3 + \frac{z^5}{2!} - \frac{z^7}{3!} + \cdots$$

13. Multiplying out gives $(1+x)^3 = 1 + 3x + 3x^2 + x^3$. Since this polynomial equals the original function for all x, it must be the Taylor series. The general term is $0 \cdot x^n$ for $n \geq 4$.

17.

$$\frac{a}{\sqrt{a^2+x^2}} = \frac{a}{a(1+\frac{x^2}{a^2})^{\frac{1}{2}}} = \left(1 + \frac{x^2}{a^2}\right)^{-\frac{1}{2}}$$
$$= 1 + \left(-\frac{1}{2}\right)\frac{x^2}{a^2} + \frac{1}{2!}\left(-\frac{1}{2}\right)\left(-\frac{3}{2}\right)\left(\frac{x^2}{a^2}\right)^2$$
$$+ \frac{1}{3!}\left(-\frac{1}{2}\right)\left(-\frac{3}{2}\right)\left(-\frac{5}{2}\right)\left(\frac{x^2}{a^2}\right)^3 + \cdots$$
$$= 1 - \frac{1}{2}\left(\frac{x}{a}\right)^2 + \frac{3}{8}\left(\frac{x}{a}\right)^4 - \frac{5}{16}\left(\frac{x}{a}\right)^6 + \cdots$$

Problems

21. The Taylor series about 0 for $y = \dfrac{1}{1-x^2}$ is

$$y = 1 + x^2 + x^4 + x^6 + \cdots.$$

The series for $y = (1 + x)^{1/4}$ is, using the binomial expansion,

$$y = 1 + \frac{1}{4}x + \frac{1}{4}\left(-\frac{3}{4}\right)\frac{x^2}{2!} + \frac{1}{4}\left(-\frac{3}{4}\right)\left(-\frac{7}{4}\right)\frac{x^3}{3!} + \cdots.$$

The series for $y = \sqrt{1 + \frac{x}{2}} = (1 + \frac{x}{2})^{1/2}$ is, again using the binomial expansion,

$$y = 1 + \frac{1}{2} \cdot \frac{x}{2} + \frac{1}{2}\left(-\frac{1}{2}\right) \cdot \frac{x^2}{8} + \frac{1}{2}\left(-\frac{1}{2}\right)\left(-\frac{3}{2}\right) \cdot \frac{x^3}{48} + \cdots.$$

Similarly for $y = \dfrac{1}{\sqrt{1-x}} = (1-x)^{-(1/2)}$,

$$y = 1 + \left(-\frac{1}{2}\right)(-x) + \left(-\frac{1}{2}\right)\left(-\frac{3}{2}\right) \cdot \frac{x^2}{2!} + \left(-\frac{1}{2}\right)\left(-\frac{3}{2}\right)\left(-\frac{5}{2}\right) \cdot \frac{-x^3}{3!} + \cdots.$$

Near 0, let's truncate these series after their x^2 terms:

$$\frac{1}{1 - x^2} \approx 1 + x^2,$$

$$(1 + x)^{1/4} \approx 1 + \frac{1}{4}x - \frac{3}{32}x^2,$$

$$\sqrt{1 + \frac{x}{2}} \approx 1 + \frac{1}{4}x - \frac{1}{32}x^2,$$

$$\frac{1}{\sqrt{1-x}} \approx 1 + \frac{1}{2}x + \frac{3}{8}x^2.$$

Thus $\frac{1}{1-x^2}$ looks like a parabola opening upward near the origin, with y-axis as the axis of symmetry, so (a) = I.

Now $\frac{1}{\sqrt{1-x}}$ has the largest positive slope ($\frac{1}{2}$), and is concave up (because the coefficient of x^2 is positive). So (d) = II.

The last two both have positive slope ($\frac{1}{4}$) and are concave down. Since $(1 + x)^{\frac{1}{4}}$ has the smallest second derivative (i.e., the most negative coefficient of x^2), (b) = IV and therefore (c) = III.

25.

$$E = kQ\left(\frac{1}{(R-1)^2} - \frac{1}{(R+1)^2}\right)$$

$$= \frac{kQ}{R^2}\left(\frac{1}{(1 - \frac{1}{R})^2} - \frac{1}{(1 + \frac{1}{R})^2}\right)$$

Since $\left|\frac{1}{R}\right| < 1$, we can expand the two terms using the binomial expansion:

$$\frac{1}{(1 - \frac{1}{R})^2} = \left(1 - \frac{1}{R}\right)^{-2}$$

$$= 1 - 2\left(-\frac{1}{R}\right) + (-2)(-3)\frac{(-\frac{1}{R})^2}{2!} + (-2)(-3)(-4)\frac{(-\frac{1}{R})^3}{3!} + \cdots$$

$$\frac{1}{(1 + \frac{1}{R})^2} = \left(1 + \frac{1}{R}\right)^{-2}$$

$$= 1 - 2\left(\frac{1}{R}\right) + (-2)(-3)\frac{(\frac{1}{R})^2}{2!} + (-2)(-3)(-4)\frac{(\frac{1}{R})^3}{3!} + \cdots$$

Substituting, we get:

$$E = \frac{kQ}{R^2}\left[1 + \frac{2}{R} + \frac{3}{R^2} + \frac{4}{R^3} + \cdots - \left(1 - \frac{2}{R} + \frac{3}{R^2} - \frac{4}{R^3} + \cdots\right)\right] \approx \frac{kQ}{R^2}\left(\frac{4}{R} + \frac{8}{R^3}\right),$$

using only the first two non-zero terms.

29. (a) Factoring the expression for $t_1 - t_2$, we get

$$\Delta t = t_1 - t_2 = \frac{2l_2}{c(1 - v^2/c^2)} - \frac{2l_1}{c\sqrt{1 - v^2/c^2}} - \frac{2l_2}{c\sqrt{1 - v^2/c^2}} + \frac{2l_1}{c(1 - v^2/c^2)}$$

$$= \frac{2(l_1 + l_2)}{c(1 - v^2/c^2)} - \frac{2(l_1 + l_2)}{c\sqrt{1 - v^2/c^2}}$$

$$= \frac{2(l_1 + l_2)}{c}\left(\frac{1}{1 - v^2/c^2} - \frac{1}{\sqrt{1 - v^2/c^2}}\right).$$

Expanding the two terms within the parentheses in terms of v^2/c^2 gives

$$\left(1 - \frac{v^2}{c^2}\right)^{-1} = 1 + \frac{v^2}{c^2} + \frac{(-1)(-2)}{2!}\left(\frac{-v^2}{c^2}\right)^2 + \frac{(-1)(-2)(-3)}{3!}\left(\frac{-v^2}{c^2}\right)^3 + \cdots$$

$$= 1 + \frac{v^2}{c^2} + \frac{v^4}{c^4} + \frac{v^6}{c^6} + \cdots$$

$$\left(1 - \frac{v^2}{c^2}\right)^{-1/2} = 1 + \frac{1}{2}\frac{v^2}{c^2} + \frac{\left(\frac{-1}{2}\right)\left(\frac{-3}{2}\right)}{2!}\left(\frac{-v^2}{c^2}\right)^2 + \frac{\left(\frac{-1}{2}\right)\left(\frac{-3}{2}\right)\left(\frac{-5}{2}\right)}{3!}\left(\frac{-v^2}{c^2}\right)^3 + \cdots$$

$$= 1 + \frac{1}{2}\frac{v^2}{c^2} + \frac{3}{8}\frac{v^4}{c^4} + \frac{5}{16}\frac{v^6}{c^6} + \cdots$$

Thus, we have

$$\Delta t = \frac{2(l_1 + l_2)}{c}\left(1 + \frac{v^2}{c^2} + \frac{v^4}{c^4} + \frac{v^6}{c^6} + \cdots - 1 - \frac{1}{2}\frac{v^2}{c^2} - \frac{3}{8}\frac{v^4}{c^4} - \frac{5}{16}\frac{v^6}{c^6} - \cdots\right)$$

$$= \frac{2(l_1 + l_2)}{c}\left(\frac{1}{2}\frac{v^2}{c^2} + \frac{5}{8}\frac{v^4}{c^4} + \frac{11}{16}\frac{v^6}{c^6} + \cdots\right)$$

$$\Delta t \approx \frac{(l_1 + l_2)}{c}\left(\frac{v^2}{c^2} + \frac{5}{4}\frac{v^4}{c^4}\right).$$

(b) For small v. we can neglect all but the first nonzero term, so

$$\Delta t \approx \frac{(l_1 + l_2)}{c} \cdot \frac{v^2}{c^2} = \frac{(l_1 + l_2)}{c^3}v^2.$$

Thus, Δt is proportional to v^2 with constant of proportionality $(l_1 + l_2)/c^3$.

33. (a) We take the left-hand Riemann sum with the formula

$$\text{Left-hand sum} = (1 + 0.9608 + 0.8521 + 0.6977 + 0.5273)(0.2) = 0.8076.$$

Similarly,

$$\text{Right-hand sum} = (0.9608 + 0.8521 + 0.6977 + 0.5273 + 0.3679)(0.2) = 0.6812.$$

(b) Since

$$e^x = 1 + x + \frac{x^2}{2!} + \frac{x^3}{3!} + \cdots,$$

$$e^{-x^2} \approx 1 + (-x^2) + \frac{(-x^2)^2}{2!} + \frac{(-x^2)^3}{3!}$$

$$= 1 - x^2 + \frac{x^4}{2} - \frac{x^6}{6}.$$

(c)

$$\int_0^1 e^{-x^2}\,dx \approx \int_0^1 \left(1 - x^2 + \frac{x^4}{2} - \frac{x^6}{6}\right)dx$$

$$= \left(x - \frac{x^3}{3} + \frac{x^5}{10} - \frac{x^7}{42}\right)\Big|_0^1 = 0.74286.$$

(d) We can improve the left and right sum values by averaging them to get 0.74439 or by increasing the number of subdivisions. We can improve on the estimate using the Taylor approximation by taking more terms.

Solutions for Section 10.4

Exercises

1. Let $f(x) = (1-x)^{1/3}$, so $f(0.5) = (0.5)^{1/3}$. The error bound in the Taylor approximation of degree 3 for $f(0.5) = 0.5^{\frac{1}{3}}$ about $x = 0$ is:

$$|E_3| = |f(0.5) - P_3(0.5)| \leq \frac{M \cdot |0.5 - 0|^4}{4!} = \frac{M(0.5)^4}{24},$$

where $|f^{(4)}(x)| \leq M$ for $0 \leq x \leq 0.5$. Now, $f^{(4)}(x) = -\frac{80}{81}(1-x)^{-(11/3)}$. By looking at the graph of $(1-x)^{-(11/3)}$, we see that $|f^{(4)}(x)|$ is maximized for x between 0 and 0.5 when $x = 0.5$. Thus,

$$|f^{(4)}| \leq \frac{80}{81}\left(\frac{1}{2}\right)^{-(11/3)} = \frac{80}{81} \cdot 2^{11/3},$$

so

$$|E_3| \leq \frac{80 \cdot 2^{11/3} \cdot (0.5)^4}{81 \cdot 24} \approx 0.033.$$

Problems

5. (a) The Taylor polynomial of degree 0 about $t = 0$ for $f(t) = e^t$ is simply $P_0(x) = 1$. Since $e^t \geq 1$ on $[0, 0.5]$, the approximation is an underestimate.

(b) Using the zero degree error bound, if $|f'(t)| \leq M$ for $0 \leq t \leq 0.5$, then

$$|E_0| \leq M \cdot |t| \leq M(0.5).$$

Since $|f'(t)| = |e^t| = e^t$ is increasing on $[0, 0.5]$,

$$|f'(t)| \leq e^{0.5} < \sqrt{4} = 2.$$

Therefore

$$|E_0| \leq (2)(0.5) = 1.$$

(Note: By looking at a graph of $f(t)$ and its 0th degree approximation, it is easy to see that the greatest error occurs when $t = 0.5$, and the error is $e^{0.5} - 1 \approx 0.65 < 1$. So our error bound works.)

9. (a) The vertical distance between the graph of $y = \cos x$ and $y = P_{10}(x)$ at $x = 6$ is no more than 4, so

$$|\text{Error in } P_{10}(6)| \leq 4.$$

Since at $x = 6$ the $\cos x$ and $P_{20}(x)$ graphs are indistinguishable in this figure, the error must be less than the smallest division we can see, which is about 0.2 so,

$$|\text{Error in } P_{20}(6)| \leq 0.2.$$

(b) The maximum error occurs at the ends of the interval, that is, at $x = -9, x = 9$. At $x = 9$, the graphs of $y = \cos x$ and $y = P_{20}(x)$ are no more than 1 apart, so

$$\left|\begin{matrix}\text{Maximum error in } P_{20}(x)\\ \text{for } -9 \leq x \leq 9\end{matrix}\right| \leq 1.$$

(c) We are looking for the largest x-interval on which the graphs of $y = \cos x$ and $y = P_{10}(x)$ are indistinguishable. This is hard to estimate accurately from the figure, though $-4 \leq x \leq 4$ certainly satisfies this condition.

13. (a)

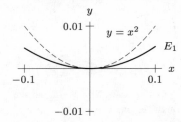

The graph of E_1 looks like a parabola. Since the graph of E_1 is sandwiched between the graph of $y = x^2$ and the x axis, we have

$$|E_1| \le x^2 \quad \text{for} \quad |x| \le 0.1.$$

(b)

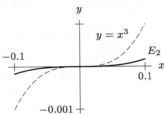

The graph of E_2 looks like a cubic, sandwiched between the graph of $y = x^3$ and the x axis, so

$$|E_2| \le x^3 \quad \text{for} \quad |x| \le 0.1.$$

(c) Using the Taylor expansion

$$e^x = 1 + x + \frac{x^2}{2!} + \frac{x^3}{3!} + \cdots$$

we see that

$$E_1 = e^x - (1 + x) = \frac{x^2}{2!} + \frac{x^3}{3!} + \frac{x^4}{4!} + \cdots.$$

Thus for small x, the $x^2/2!$ term dominates, so

$$E_1 \approx \frac{x^2}{2!},$$

and so E_1 is approximately a quadratic.

Similarly

$$E_2 = e^x - (1 + x + \frac{x^2}{2}) = \frac{x^3}{3!} + \frac{x^4}{4!} + \cdots.$$

Thus for small x, the $x^3/3!$ term dominates, so

$$E_2 \approx \frac{x^3}{3!}$$

and so E_2 is approximately a cubic.

Solutions for Section 10.5

Exercises

1. No, a Fourier series has terms of the form $\cos nx$, not $\cos^n x$.

5.

$$a_0 = \frac{1}{2\pi} \int_{-\pi}^{\pi} f(x)\, dx = \frac{1}{2\pi} \left[\int_{-\pi}^{0} -1\, dx + \int_{0}^{\pi} 1\, dx \right] = 0$$

$$a_1 = \frac{1}{\pi} \int_{-\pi}^{\pi} f(x) \cos x\, dx = \frac{1}{\pi} \left[\int_{-\pi}^{0} -\cos x\, dx + \int_{0}^{\pi} \cos x\, dx \right]$$

$$= \frac{1}{\pi} \left[-\sin x \Big|_{-\pi}^{0} + \sin x \Big|_{0}^{\pi} \right] = 0.$$

Similarly, a_2 and a_3 are both 0.

(In fact, notice $f(x) \cos nx$ is an odd function, so $\int_{-\pi}^{\pi} f(x) \cos nx = 0$.)

$$b_1 = \frac{1}{\pi} \int_{-\pi}^{\pi} f(x) \sin x\, dx = \frac{1}{\pi} \left[\int_{-\pi}^{0} -\sin x\, dx + \int_{0}^{\pi} \sin x\, dx \right]$$

$$= \frac{1}{\pi} \left[\cos x \Big|_{-\pi}^{0} + (-\cos x) \Big|_{0}^{\pi} \right] = \frac{4}{\pi}.$$

$$b_2 = \frac{1}{\pi} \int_{-\pi}^{\pi} f(x) \sin 2x \, dx = \frac{1}{\pi} \left[\int_{-\pi}^{0} -\sin 2x \, dx + \int_{0}^{\pi} \sin 2x \, dx \right]$$

$$= \frac{1}{\pi} \left[\frac{1}{2} \cos 2x \Big|_{-\pi}^{0} + (-\frac{1}{2} \cos 2x) \Big|_{0}^{\pi} \right] = 0.$$

$$b_3 = \frac{1}{\pi} \int_{-\pi}^{\pi} f(x) \sin 3x \, dx = \frac{1}{\pi} \left[\int_{-\pi}^{0} -\sin 3x \, dx + \int_{0}^{\pi} \sin 3x \, dx \right]$$

$$= \frac{1}{\pi} \left[\frac{1}{3} \cos 3x \Big|_{-\pi}^{0} + (-\frac{1}{3} \cos 3x) \Big|_{0}^{\pi} \right] = \frac{4}{3\pi}.$$

Thus, $F_1(x) = F_2(x) = \frac{4}{\pi} \sin x$ and $F_3(x) = \frac{4}{\pi} \sin x + \frac{4}{3\pi} \sin 3x$.

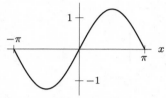

$$F_1(x) = F_2(x) = \frac{4}{\pi} \sin x$$

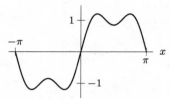

$$F_3(x) = \frac{4}{\pi} \sin x + \frac{4}{3\pi} \sin 3x$$

9.

$$a_0 = \frac{1}{2\pi} \int_{-\pi}^{\pi} h(x) \, dx = \frac{1}{2\pi} \int_{0}^{\pi} x \, dx = \frac{\pi}{4}$$

As in Problem 10, we use the integral table (III-15 and III-16) to find formulas for a_n and b_n.

$$a_n = \frac{1}{\pi} \int_{-\pi}^{\pi} h(x) \cos(nx) \, dx = \frac{1}{\pi} \int_{0}^{\pi} x \cos nx \, dx = \frac{1}{\pi} \left(\frac{x}{n} \sin(nx) + \frac{1}{n^2} \cos(nx) \right) \Big|_{0}^{\pi}$$

$$= \frac{1}{\pi} \left(\frac{1}{n^2} \cos(n\pi) - \frac{1}{n^2} \right)$$

$$= \frac{1}{n^2 \pi} \left(\cos(n\pi) - 1 \right).$$

Note that since $\cos(n\pi) = (-1)^n$, $a_n = 0$ if n is even and $a_n = -\frac{2}{n^2\pi}$ if n is odd.

$$b_n = \frac{1}{\pi} \int_{-\pi}^{\pi} h(x) \cos(nx) \, dx = \frac{1}{\pi} \int_{0}^{\pi} x \sin x \, dx$$

$$= \frac{1}{\pi} \left(-\frac{x}{n} \cos(nx) + \frac{1}{n^2} \sin(nx) \right) \Big|_{0}^{\pi}$$

$$= \frac{1}{\pi} \left(-\frac{\pi}{n} \cos(n\pi) \right)$$

$$= -\frac{1}{n} \cos(n\pi)$$

$$= \frac{1}{n} (-1)^{n+1} \quad \text{if } n \geq 1$$

We have that the nth Fourier polynomial for h (for $n \geq 1$) is

$$H_n(x) = \frac{\pi}{4} + \sum_{i=1}^{n} \left(\frac{1}{i^2 \pi} \left(\cos(i\pi) - 1 \right) \cdot \cos(ix) + \frac{(-1)^{i+1} \sin(ix)}{i} \right).$$

This can also be written as

$$H_n(x) = \frac{\pi}{4} + \sum_{i=1}^{n} \frac{(-1)^{i+1} \sin(ix)}{i} + \sum_{i=1}^{\left[\frac{n}{2}\right]} \frac{-2}{(2i-1)^2 \pi} \cos((2i-1)x)$$

where $\left[\frac{n}{2}\right]$ denotes the biggest integer smaller than or equal to $\frac{n}{2}$. In particular, we have the graphs in Figure 10.1.

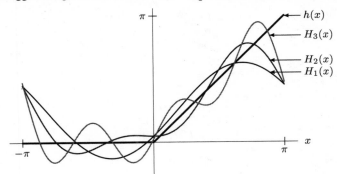

Figure 10.1

Problems

13. Since the period is 2, we make the substitution $t = \pi x - \pi$. Thus, $x = \frac{t+\pi}{\pi}$. We find the Fourier coefficients. Notice that all of the integrals are the same as in Problem 12 except for an extra factor of 2. Thus, $a_0 = 1$, $a_n = 0$, and $b_n = \frac{4}{\pi n}(-1)^{n+1}$, so:

$$G_4(t) = 1 + \frac{4}{\pi} \sin t - \frac{2}{\pi} \sin 2t + \frac{4}{3\pi} \sin 3t - \frac{1}{\pi} \sin 4t.$$

Again, we substitute back in to get a Fourier polynomial in terms of x:

$$F_4(x) = 1 + \frac{4}{\pi} \sin(\pi x - \pi) - \frac{2}{\pi} \sin(2\pi x - 2\pi)$$

$$+ \frac{4}{3\pi} \sin(3\pi x - 3\pi) - \frac{1}{\pi} \sin(4\pi x - 4\pi)$$

$$= 1 - \frac{4}{\pi} \sin(\pi x) - \frac{2}{\pi} \sin(2\pi x) - \frac{4}{3\pi} \sin(3\pi x) - \frac{1}{\pi} \sin(4\pi x).$$

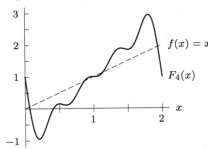

Notice in this case, the terms in our series are $\sin(n\pi x)$, not $\sin(2\pi n x)$, as in Problem 12. In general, the terms will be $\sin(n\frac{2\pi}{b}x)$, where b is the period.

17. Since each square in the graph has area $\left(\frac{\pi}{4}\right) \cdot (0.2)$,

$$a_0 = \frac{1}{2\pi} \int_{-\pi}^{\pi} f(x)\, dx$$

$$= \frac{1}{2\pi} \cdot \left(\frac{\pi}{4}\right) \cdot (0.2) \, [\text{Number of squares under graph above } x\text{-axis}$$

$$- \text{Number of squares above graph below } x \text{ axis}]$$

$$\approx \frac{1}{2\pi} \cdot \left(\frac{\pi}{4}\right) \cdot (0.2) \cdot [13 + 11 - 14] = 0.25.$$

Approximate the Fourier coefficients using Riemann sums.

$$a_1 = \frac{1}{\pi} \int_{-\pi}^{\pi} f(x) \cos x \, dx$$

$$\approx \frac{1}{\pi} \left[f(-\pi) \cos(-\pi) + f\left(-\frac{\pi}{2}\right) \cos\left(-\frac{\pi}{2}\right) + f(0) \cos(0) + f\left(\frac{\pi}{2}\right) \cos\left(\frac{\pi}{2}\right) \right] \cdot \frac{\pi}{2}$$

$$= \frac{1}{\pi} [(0.92)(-1) + (1)(0) + (-1.7)(1) + (0.7)(0)] \cdot \frac{\pi}{2}$$

$$= -1.31$$

Similarly for b_1:

$$b_1 = \frac{1}{\pi} \int_{-\pi}^{\pi} f(x) \sin x \, dx$$

$$\approx \frac{1}{\pi} \left[f(-\pi) \sin(-\pi) + f\left(-\frac{\pi}{2}\right) \sin\left(-\frac{\pi}{2}\right) + f(0) \sin(0) + f\left(\frac{\pi}{2}\right) \sin\left(\frac{\pi}{2}\right) \right] \cdot \frac{\pi}{2}$$

$$= \frac{1}{\pi} [(0.92)(0) + (1)(-1) + (-1.7)(0) + (0.7)(1)] \cdot \frac{\pi}{2}$$

$$= -0.15.$$

So our first Fourier approximation is

$$F_1(x) = 0.25 - 1.31 \cos x - 0.15 \sin x.$$

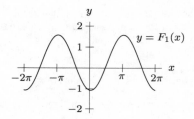

Similarly for a_2:

$$a_2 = \frac{1}{\pi} \int_{-\pi}^{\pi} f(x) \cos 2x \, dx$$

$$\approx \frac{1}{\pi} \left[f(-\pi) \cos(-2\pi) + f\left(-\frac{\pi}{2}\right) \cos(-\pi) + f(0) \cos(0) + f\left(\frac{\pi}{2}\right) \cos(-\pi) \right] \cdot \frac{\pi}{2}$$

$$= \frac{1}{\pi} [(0.92)(1) + (1)(-1) + (-1.7)(1) + (0.7)(-1)] \cdot \frac{\pi}{2}$$

$$= -1.24$$

Similarly for b_2:

$$b_2 = \frac{1}{\pi} \int_{-\pi}^{\pi} f(x) \sin 2x \, dx$$

$$\approx \frac{1}{\pi} \left[f(-\pi) \sin(-2\pi) + f\left(-\frac{\pi}{2}\right) \sin(-\pi) + f(0) \sin(0) + f\left(\frac{\pi}{2}\right) \sin(-\pi) \right] \cdot \frac{\pi}{2}$$

$$= \frac{1}{\pi} [(0.92)(0) + (1)(0) + (-1.7)(0) + (0.7)(0)] \cdot \frac{\pi}{2}$$

$$= 0.$$

So our second Fourier approximation is

$$F_2(x) = 0.25 - 1.31 \cos x - 0.15 \sin x - 1.24 \cos 2x.$$

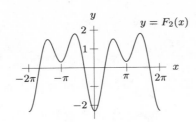

As you can see from comparing our graphs of F_1 and F_2 to the original, our estimates of the Fourier coefficients are not very accurate.

There are other methods of estimating the Fourier coefficients such as taking other Riemann sums, using Simpson's rule, and using the trapezoid rule. With each method, the greater the number of subdivisions, the more accurate the estimates of the Fourier coefficients.

The actual function graphed in the problem was

$$y = \frac{1}{4} - 1.3 \cos x - \frac{\sin(\frac{3}{5})}{\pi} \sin x - \frac{2}{\pi} \cos 2x - \frac{\cos 1}{3\pi} \sin 2x$$

$$= 0.25 - 1.3 \cos x - 0.18 \sin x - 0.63 \cos 2x - 0.057 \sin 2x.$$

21. (a)

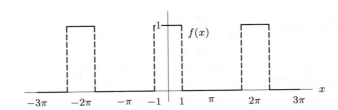

The energy of the pulse train f is

$$E = \frac{1}{\pi} \int_{-\pi}^{\pi} (f(x))^2 \, dx = \frac{1}{\pi} \int_{-1}^{1} 1^2 = \frac{1}{\pi}(1 - (-1)) = \frac{2}{\pi}.$$

Next, find the Fourier coefficients:

$$a_0 = \text{average value of } f \text{ on } [-\pi, \pi] = \frac{1}{2\pi}(\text{ Area}) = \frac{1}{2\pi}(2) = \frac{1}{\pi},$$

$$a_k = \frac{1}{\pi} \int_{-\pi}^{\pi} f(x) \cos kx \, dx = \frac{1}{\pi} \int_{-1}^{1} \cos kx \, dx = \frac{1}{k\pi} \sin kx \Big|_{-1}^{1}$$

$$= \frac{1}{k\pi}(\sin k - \sin(-k)) = \frac{1}{k\pi}(2 \sin k),$$

$$b_k = \frac{1}{\pi} \int_{-\pi}^{\pi} f(x) \sin kx \, dx = \frac{1}{\pi} \int_{-1}^{1} \sin kx \, dx = -\frac{1}{k\pi} \cos kx \Big|_{-1}^{1}$$

$$= -\frac{1}{k\pi}(\cos k - \cos(-k)) = \frac{1}{k\pi}(0) = 0.$$

The energy of f contained in the constant term is

$$A_0^2 = 2a_0^2 = 2\left(\frac{1}{\pi}\right)^2 = \frac{2}{\pi^2}$$

which is

$$\frac{A_0^2}{E} = \frac{2/\pi^2}{2/\pi} = \frac{1}{\pi} \approx 0.3183 = 31.83\% \quad \text{of the total.}$$

The fraction of energy contained in the first harmonic is

$$\frac{A_1^2}{E} = \frac{a_1^2}{E} = \frac{\left(\frac{2\sin 1}{\pi}\right)^2}{\frac{2}{\pi}} \approx 0.4508 = 45.08\%.$$

The fraction of energy contained in both the constant term and the first harmonic together is

$$\frac{A_0^2}{E} + \frac{A_1^2}{E} \approx 0.7691 = 76.91\%.$$

(b) The fraction of energy contained in the second harmonic is

$$\frac{A_2^2}{E} = \frac{a_2^2}{E} = \frac{\left(\frac{\sin 2}{\pi}\right)^2}{\frac{2}{\pi}} \approx 0.1316 = 13.16\%$$

so the fraction of energy contained in the constant term and first two harmonics is

$$\frac{A_0^2}{E} + \frac{A_1^2}{E} + \frac{A_2^2}{E} \approx 0.7691 + 0.1316 = 0.9007 = 90.07\%.$$

Therefore, the constant term and the first two harmonics are needed to capture 90% of the energy of f.

(c)

$$F_3(x) = \frac{1}{\pi} + \frac{2\sin 1}{\pi}\cos x + \frac{\sin 2}{\pi}\cos 2x + \frac{2\sin 3}{3\pi}\cos 3x$$

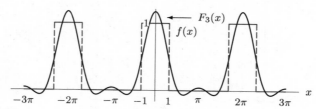

25. The easiest way to do this is to use Problem 24.

$$\int_{-\pi}^{\pi} \sin^2 mx\, dx = \int_{-\pi}^{\pi}(1 - \cos^2 mx)\, dx = \int_{-\pi}^{\pi} dx - \int_{-\pi}^{\pi}\cos^2 mx\, dx$$
$$= 2\pi - \pi \quad \text{using Problem 24}$$
$$= \pi.$$

Solutions for Chapter 10 Review

Exercises

1. $e^x \approx 1 + e(x - 1) + \frac{e}{2}(x - 1)^2$

5. $f'(x) = 3x^2 + 14x - 5$, $f''(x) = 6x + 14$, $f'''(x) = 6$. The Taylor polynomial about $x = 1$ is

$$P_3(x) = 4 + \frac{12}{1!}(x - 1) + \frac{20}{2!}(x - 1)^2 + \frac{6}{3!}(x - 1)^3$$
$$= 4 + 12(x - 1) + 10(x - 1)^2 + (x - 1)^3.$$

Notice that if you multiply out and collect terms in $P_3(x)$, you will get $f(x)$ back.

9. Substituting $y = -4z^2$ into $\dfrac{1}{1 + y} = 1 - y + y^2 - y^3 + \cdots$ gives

$$\frac{1}{1 - 4z^2} = 1 + 4z^2 + 16z^4 + 64z^6 + \cdots.$$

Problems

13. Infinite geometric series with $a = 1$, $x = -1/3$, so

$$\text{Sum} = \frac{1}{1 - (-1/3)} = \frac{3}{4}.$$

17. Factoring out a 3, we see

$$3\left(1 + 1 + \frac{1}{2!} + \frac{1}{3!} + \frac{1}{4!} + \frac{1}{5!} + \cdots\right) = 3e^1 = 3e.$$

21. The graph in Figure 10.2 suggests that the Taylor polynomials converge to $f(x) = \dfrac{1}{1+x}$ on the interval $(-1, 1)$. The Taylor expansion is

$$f(x) = \frac{1}{1+x} = 1 - x + x^2 - x^3 + x^4 - \cdots,$$

so the ratio test gives

$$\lim_{n \to \infty} \frac{|a_{n+1}|}{|a_n|} = \lim_{n \to \infty} \frac{|(-1)^{n+1} x^{n+1}|}{|(-1)^n x^n|} = |x|.$$

Thus, the series converges if $|x| < 1$; that is $-1 < x < 1$.

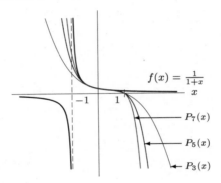

$$f(x) = \frac{1}{1+x}$$

$$x$$

$$P_7(x)$$

$$P_5(x)$$

$$P_3(x)$$

Figure 10.2

25. (a) Since $\sqrt{4 - x^2} = 2\sqrt{1 - x^2/4}$, we use the Binomial expansion

$$\sqrt{4 - x^2} \approx 2\left(1 + \frac{1}{2}\left(-\frac{x^2}{4}\right) + \frac{1}{2!}\left(\frac{1}{2}\right)\left(-\frac{1}{2}\right)\left(-\frac{x^2}{4}\right)^2\right)$$

$$= 2\left(1 - \frac{x^2}{8} - \frac{x^4}{128}\right) = 2 - \frac{x^2}{4} - \frac{x^4}{64}.$$

(b) Substituting the Taylor series in the integral gives

$$\int_0^1 \sqrt{4 - x^2}\, dx \approx \int_0^1 \left(2 - \frac{x^2}{4} - \frac{x^4}{64}\right) dx = 2x - \frac{x^3}{12} - \frac{x^5}{320}\bigg|_0^1 = 1.9135.$$

(c) Since $x = 2\sin t$, we have $dx = 2\cos t\, dt$; in addition $t = 0$ when $x = 0$ and $t = \pi/6$ when $x = 1$. Thus

$$\int_0^1 \sqrt{4 - x^2}\, dx = \int_0^{\pi/6} \sqrt{4 - 4\sin^2 t} \cdot 2\cos t\, dt$$

$$= \int_0^{\pi/6} 2 \cdot 2\sqrt{1 - \sin^2 t}\cos t\, dt = 4\int_0^{\pi/6} \cos^2 t\, dt.$$

Using the table of integrals, we find

$$4\int_0^{\pi/6} \cos^2 t\, dt = 4 \cdot \frac{1}{2}(\cos t \sin t + t)\bigg|_0^{\pi/6} = 2\left(\cos\frac{\pi}{6}\sin\frac{\pi}{6} + \frac{\pi}{6}\right) = \frac{\sqrt{3}}{2} + \frac{\pi}{3}.$$

(d) Using a calculator, $(\sqrt{3}/3) + (\pi/3) = 1.9132$, so the answers to parts (b) and (c) agree to three decimal places.

29. (a) To find when V takes on its minimum values, set $\frac{dV}{dr} = 0$. So

$$-V_0 \frac{d}{dr}\left(2\left(\frac{r_0}{r}\right)^6 - \left(\frac{r_0}{r}\right)^{12}\right) = 0$$

$$-V_0\left(-12r_0^6 r^{-7} + 12r_0^{12}r^{-13}\right) = 0$$

$$12r_0^6 r^{-7} = 12r_0^{12}r^{-13}$$

$$r_0^6 = r^6$$

$$r = r_0.$$

Rewriting $V'(r)$ as $\frac{12r_0^6 V_0}{r^7}\left(1 - \left(\frac{r_0}{r}\right)^6\right)$, we see that $V'(r) > 0$ for $r > r_0$ and $V'(r) < 0$ for $r < r_0$. Thus, $V = -V_0(2(1)^6 - (1)^{12}) = -V_0$ is a minimum.

(Note: We discard the negative root $-r_0$ since the distance r must be positive.)

(b)

$$V(r) = -V_0\left(2\left(\frac{r_0}{r}\right)^6 - \left(\frac{r_0}{r}\right)^{12}\right) \qquad V(r_0) = -V_0$$

$$V'(r) = -V_0(-12r_0^6 r^{-7} + 12r_0^{12}r^{-13}) \qquad V'(r_0) = 0$$

$$V''(r) = -V_0(84r_0^6 r^{-8} - 156r_0^{12}r^{-14}) \qquad V''(r_0) = 72V_0 r_0^{-2}$$

The Taylor series is thus:

$$V(r) = -V_0 + 72V_0 r_0^{-2} \cdot (r - r_0)^2 \cdot \frac{1}{2} + \cdots$$

(c) The difference between V and its minimum value $-V_0$ is

$$V - (-V_0) = 36V_0 \frac{(r - r_0)^2}{r_0^2} + \cdots$$

which is approximately proportional to $(r - r_0)^2$ since terms containing higher powers of $(r - r_0)$ have relatively small values for r near r_0.

(d) From part (a) we know that $dV/dr = 0$ when $r = r_0$, hence $F = 0$ when $r = r_0$. Since, if we discard powers of $(r - r_0)$ higher than the second,

$$V(r) \approx -V_0\left(1 - 36\frac{(r - r_0)^2}{r_0^2}\right)$$

giving

$$F = -\frac{dV}{dr} \approx 72 \cdot \frac{r - r_0}{r_0^2}(-V_0) = -72V_0\frac{r - r_0}{r_0^2}.$$

So F is approximately proportional to $(r - r_0)$.

33. (a) Notice $g'(0) = 0$ because g has a critical point at $x = 0$. So, for $n \geq 2$,

$$g(x) \approx P_n(x) = g(0) + \frac{g''(0)}{2!}x^2 + \frac{g'''(0)}{3!}x^3 + \cdots + \frac{g^{(n)}(0)}{n!}x^n.$$

(b) The Second Derivative test says that if $g''(0) > 0$, then 0 is a local minimum and if $g''(0) < 0$, 0 is a local maximum.

(c) Let $n = 2$. Then $P_2(x) = g(0) + \frac{g''(0)}{2!}x^2$. So, for x near 0,

$$g(x) - g(0) \approx \frac{g''(0)}{2!}x^2.$$

If $g''(0) > 0$, then $g(x) - g(0) \geq 0$, as long as x stays near 0. In other words, there exists a small interval around $x = 0$ such that for any x in this interval $g(x) \geq g(0)$. So $g(0)$ is a local minimum.

The case when $g''(0) < 0$ is treated similarly; then $g(0)$ is a local maximum.

37. (a) Expand $f(x)$ into its Fourier series:

$$f(x) = a_0 + a_1 \cos x + a_2 \cos 2x + a_3 \cos 3x + \cdots + a_k \cos kx + \cdots$$
$$+ b_1 \sin x + b_2 \sin 2x + b_3 \sin 3x + \cdots + b_k \sin kx + \cdots$$

Then differentiate term-by-term:

$$f'(x) = -a_1 \sin x - 2a_2 \sin 2x - 3a_3 \sin 3x - \cdots - ka_k \sin kx - \cdots$$
$$+ b_1 \cos x + 2b_2 \cos 2x + 3b_3 \cos 3x + \cdots + kb_k \cos kx + \cdots$$

Regroup terms:

$$f'(x) = + b_1 \cos x + 2b_2 \cos 2x + 3b_3 \cos 3x + \cdots + kb_k \cos kx + \cdots$$
$$- a_1 \sin x - 2a_2 \sin 2x - 3a_3 \sin 3x - \cdots - ka_k \sin kx - \cdots$$

which forms a Fourier series for the derivative $f'(x)$. The Fourier coefficient of $\cos kx$ is kb_k and the Fourier coefficient of $\sin kx$ is $-ka_k$. Note that there is no constant term as you would expect from the formula ka_k with $k = 0$. Note also that if the kth harmonic f is absent, so is that of f'.

(b) If the amplitude of the kth harmonic of f is

$$A_k = \sqrt{a_k^2 + b_k^2}, \quad k \geq 1,$$

then the amplitude of the kth harmonic of f' is

$$\sqrt{(kb_k)^2 + (-ka_k)^2} = \sqrt{k^2(b_k^2 + a_k^2)} = k\sqrt{a_k^2 + b_k^2} = kA_k.$$

(c) The energy of the kth harmonic of f' is k^2 times the energy of the kth harmonic of f.

CAS Challenge Problems

41. (a) The Taylor polynomials of degree 7 are

$$\text{For } \sin x, \quad P_7(x) = x - \frac{x^3}{6} + \frac{x^5}{120} - \frac{x^7}{5040}$$

$$\text{For } \sin x \cos x, \quad Q_7(x) = x - \frac{2x^3}{3} + \frac{2x^5}{15} - \frac{4x^7}{315}$$

(b) The coefficient of x^3 in $Q_7(x)$ is $-2/3$, and the coefficient of x^3 in $P_7(x)$ is $-1/6$, so the ratio is

$$\frac{-2/3}{-1/6} = 4.$$

The corresponding ratios for x^5 and x^7 are

$$\frac{2/15}{1/120} = 16 \quad \text{and} \quad \frac{-4/315}{-1/5040} = 64.$$

(c) It appears that the ratio is always a power of 2. For x^3, it is $4 = 2^2$; for x^5, it is $16 = 2^4$; for x^7, it is $64 = 2^6$. This suggests that in general, for the coefficient of x^n, it is 2^{n-1}.

(d) From the identity $\sin(2x) = 2 \sin x \cos x$, we expect that $P_7(2x) = 2Q_7(x)$. So, if a_n is the coefficient of x^n in $P_7(x)$, and if b_n is the coefficient of x^n in $Q_7(x)$, then, since the x^n terms $P_7(2x)$ and $2Q_7(x)$ must be equal, we have

$$a_n(2x)^n = 2b_n x^n.$$

Dividing both sides by x^n and combining the powers of 2, this gives the pattern we observed. For $a_n \neq 0$,

$$\frac{b_n}{a_n} = 2^{n-1}.$$

CHECK YOUR UNDERSTANDING

1. False. For example, both $f(x) = x^2$ and $g(x) = x^2 + x^3$ have $P_2(x) = x^2$.

5. False. The Taylor series for $\sin x$ about $x = \pi$ is calculated by taking derivatives and using the formula

$$f(a) + f'(a)(x - a) + \frac{f''(a)}{2!}(x - a)^2 + \cdots.$$

The series for $\sin x$ about $x = \pi$ turns out to be

$$-(x - \pi) + \frac{(x - \pi)^3}{3!} - \frac{(x - \pi)^5}{5!} + \cdots.$$

9. False. The derivative of $f(x)g(x)$ is not $f'(x)g'(x)$. If this statement were true, the Taylor series for $(\cos x)(\sin x)$ would have all zero terms.

13. True. For large x, the graph of $P_{10}(x)$ looks like the graph of its highest powered term, $x^{10}/10!$. But e^x grows faster than any power, so e^x gets further and further away from $x^{10}/10! \approx P_{10}(x)$.

17. True. Since f is even, $f(x) \sin(mx)$ is odd for any m, so

$$b_m = \frac{1}{\pi} \int_{-\pi}^{\pi} f(x) \sin x(mx) \, dx = 0.$$

21. False. The quadratic approximation to $f_1(x)f_2(x)$ near $x = 0$ is

$$f_1(0)f_2(0) + (f_1'(0)f_2(0) + f_1(0)f_2'(0))x + \frac{f_1''(0)f_2(0) + 2f_1'(0)f_2'(0) + f_1(0)f_2''(0)}{2}x^2.$$

On the other hand, we have

$$L_1(x) = f_1(0) + f_1'(0)x, \quad L_2(x) = f_2(0) + f_2'(0)x,$$

so

$$L_1(x)L_2(x) = (f_1(0) + f_1'(0)x)(f_2(0) + f_2'(0)x) = f_1(0)f_2(0) + (f_1'(0)f_2(0) + f_2'(0)f_1(0))x + f_1'(0)f_2'(0)x^2.$$

The first two terms of the right side agree with the quadratic approximation to $f_1(x)f_2(x)$ near $x = 0$, but the term of degree 2 does not.

For example, the linear approximation to e^x is $1 + x$, but the quadratic approximation to $(e^x)^2 = e^{2x}$ is $1 + 2x + 2x^2$, not $(1 + x)^2 = 1 + 2x + x^2$.

CHAPTER ELEVEN

Solutions for Section 11.1

Exercises

1. **(a)** (III) An island can only sustain the population up to a certain size. The population will grow until it reaches this limiting value.
   (b) (V) The ingot will get hot and then cool off, so the temperature will increase and then decrease.
   (c) (I) The speed of the car is constant, and then decreases linearly when the breaks are applied uniformly.
   (d) (II) Carbon-14 decays exponentially.
   (e) (IV) Tree pollen is seasonal, and therefore cyclical.

5. If y satisfies the differential equation, then we must have

$$\frac{d\,(5 + 3e^{kx})}{dx} = 10 - 2(5 + 3e^{kx})$$

$$3ke^{kx} = 10 - 10 - 6e^{kx}$$

$$3ke^{kx} = -6e^{kx}$$

$$k = -2.$$

So, if $k = -2$ the formula for y solves the differential equation.

9. If $y = \sin 2t$, then $\frac{dy}{dt} = 2\cos 2t$, and $\frac{d^2 y}{dt^2} = -4\sin 2t$.
   Thus $\frac{d^2 y}{dt^2} + 4y = -4\sin 2t + 4\sin 2t = 0$.

Problems

13. **(a)** $P = \frac{1}{1+e^{-t}} = (1 + e^{-t})^{-1}$
   $\frac{dP}{dt} = -(1 + e^{-t})^{-2}(-e^{-t}) = \frac{e^{-t}}{(1+e^{-t})^2}$.
   Then $P(1 - P) = \frac{1}{1+e^{-t}}\left(1 - \frac{1}{1+e^{-t}}\right) = \left(\frac{1}{1+e^{-t}}\right)\left(\frac{e^{-t}}{1+e^{-t}}\right) = \frac{e^{-t}}{(1+e^{-t})^2} = \frac{dP}{dt}$.
   (b) As t tends to ∞, e^{-t} goes to 0. Thus $\lim\limits_{t \to \infty} \frac{1}{1+e^{-t}} = 1$.

Solutions for Section 11.2

Exercises

1. There are many possible answers. One possibility is shown in Figures 11.1 and 11.2.

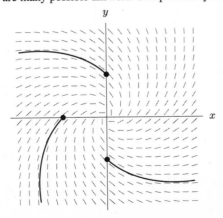

Figure 11.1

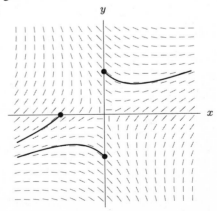

Figure 11.2

Problems

5. (a)

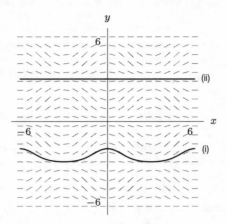

Figure 11.3

(b) We can see that the slope lines are horizontal when y is an integer multiple of π. We conclude from Figure 11.3 that the solution is $y = n\pi$ in this case.

To check this, we note that if $y = n\pi$, then $(\sin x)(\sin y) = (\sin x)(\sin n\pi) = 0 = y'$. Thus $y = n\pi$ is a solution to $y' = (\sin x)(\sin y)$, and it passes through $(0, n\pi)$.

9. (a) II   (b) VI   (c) IV   (d) I   (e) III   (f) V

Solutions for Section 11.3

Exercises

1. (a)

Table 11.1   *Euler's method for*
$y' = x + y$ *with* $y(0) = 1$

x	y	$\Delta y = (\text{slope})\Delta x$
0	1	$0.1 = (1)(0.1)$
0.1	1.1	$0.12 = (1.2)(0.1)$
0.2	1.22	$0.142 = (1.42)(0.1)$
0.3	1.362	$0.1662 = (1.662)(0.1)$
0.4	1.5282	

So $y(0.4) \approx 1.5282$.

(b)

Table 11.2   *Euler's method for*
$y' = x + y$ *with* $y(-1) = 0$

x	y	$\Delta y = (\text{slope})\Delta x$
-1	0	$-0.1 = (-1)(0.1)$
-0.9	-0.1	$-0.1 = (-1)(0.1)$
-0.8	-0.2	$-0.1 = (-1)(0.1)$
-0.7	-0.3	
$\vdots$	$\vdots$	Notice that y
0	-1	decreases by 0.1
$\vdots$	$\vdots$	for every step
0.4	-1.4	

So $y(0.4) = -1.4$. (This answer is exact.)

Problems

5. (a) (i)

Table 11.3 *Euler's method for*
$y' = (\sin x)(\sin y)$, *starting at* $(0, 2)$

x	y	$\Delta y =(\text{slope})\Delta x$
0	2	$0 = (\sin 0)(\sin 2)(0.1)$
0.1	2	$0.009 = (\sin 0.1)(\sin 2)(0.1)$
0.2	2.009	$0.018 = (\sin 0.2)(\sin 2.009)(0.1)$
0.3	2.027	

(ii)

Table 11.4 *Euler's method for*
$y' = (\sin x)(\sin y)$, *starting at*
$(0, \pi)$

x	y	$\Delta y =(\text{slope})\Delta x$
0	π	$0 = (\sin 0)(\sin \pi)(0.1)$
0.1	π	$0 = (\sin 0.1)(\sin \pi)(0.1)$
0.2	π	$0 = (\sin 0.2)(\sin \pi)(0.1)$
0.3	π	

(b) The slope field shows that the slope of the solution curve through $(0, \pi)$ is always 0. Thus the solution curve is the horizontal line with equation $y = \pi$.

9. (a) Using one step, $\frac{\Delta B}{\Delta t} = 0.05$, so $\Delta B = \left(\frac{\Delta B}{\Delta t}\right) \Delta t = 50$. Therefore we get an approximation of $B \approx 1050$ after one year.

(b) With two steps, $\Delta t = 0.5$ and we have

Table 11.5

t	B	$\Delta B = (0.05B)\Delta t$
0	1000	25
0.5	1025	25.63
1.0	1050.63	

(c) Keeping track to the nearest hundredth with $\Delta t = 0.25$, we have

Table 11.6

t	B	$\Delta B = (0.05B)\Delta t$
0	1000	12.5
0.25	1012.5	12.66
0.5	1025.16	12.81
0.75	1037.97	12.97
1	1050.94	

(d) In part (a), we get our approximation by making a single increment, ΔB, where ΔB is just $0.05B$. If we think in terms of interest, ΔB is just like getting one end of the year interest payment. Since ΔB is 0.05 times the balance B, it is like getting 5% interest at the end of the year.

(e) Part (b) is equivalent to computing the final amount in an account that begins with $1000 and earns 5% interest compounded twice annually. Each step is like computing the interest after 6 months. When $t = 0.5$, for example, the interest is $\Delta B = (0.05B) \cdot \frac{1}{2}$, and we add this to $1000 to get the new balance.

Similarly, part (c) is equivalent to the final amount in an account that has an initial balance of $1000 and earns 5% interest compounded quarterly.

Solutions for Section 11.4

Exercises

1. $\frac{dP}{dt} = 0.02P$ implies that $\frac{dP}{P} = 0.02\, dt$.

$\int \frac{dP}{P} = \int 0.02\, dt$ implies that $\ln |P| = 0.02t + C$.

$|P| = e^{0.02t+C}$ implies that $P = Ae^{0.02t}$, where $A = \pm e^C$.
We are given $P(0) = 20$. Therefore, $P(0) = Ae^{(0.02)\cdot 0} = A = 20$. So the solution is $P = 20e^{0.02t}$.

5. Separating variables and integrating both sides gives

$$\int \frac{1}{L}\, dL = \frac{1}{2} \int dp$$

or

$$\ln |L| = \frac{1}{2}p + C.$$

This can be written

$$L(p) = \pm e^{(1/2)p+C} = Ae^{p/2}.$$

The initial condition $L(0) = 100$ gives $100 = A$, so

$$L(p) = 100e^{p/2}.$$

9. $\frac{1}{z}\frac{dz}{dt} = 5$ implies $\frac{dz}{z} = 5\, dt$.
Integrating and moving terms, we have $z = Ae^{5t}$. Using the fact that $z(1) = 5$, we have $z(1) = Ae^5 = 5$, so $A = \frac{5}{e^5}$.
Therefore, $z = \frac{5}{e^5}e^{5t} = 5e^{5t-5}$.

13. $\frac{dP}{dt} = P + 4$ implies that $\frac{dP}{P+4} = dt$.

$\int \frac{dP}{P+4} = \int dt$ implies that $\ln |P + 4| = t + C$.

$P + 4 = Ae^t$ implies that $P = Ae^t - 4$. $P = 100$ when $t = 0$, so $P(0) = Ae^0 - 4 = 100$, and $A = 104$. Therefore $P = 104e^t - 4$.

17. We know that the general solution to a differential equation of the form

$$\frac{dy}{dt} = k(y - A)$$

is

$$y = Ce^{kt} + A.$$

Thus, in our case, we get

$$y = Ce^{t/2} + 200.$$

We know that at $t = 0$ we have $y = 50$, so solving for C we get

$$y = Ce^{t/2} + 200$$
$$50 = Ce^{0/2} + 200$$
$$-150 = Ce^0$$
$$C = -150.$$

Thus we get

$$y = 200 - 150e^{t/2}.$$

21. $\frac{dz}{dt} = te^z$ implies $e^{-z}dz = tdt$ implies $\int e^{-z}\, dz = \int t\, dt$ implies $-e^{-z} = \frac{t^2}{2} + C$.
Since the solution passes through the origin, $z = 0$ when $t = 0$, we must have $-e^{-0} = \frac{0}{2} + C$, so $C = -1$. Thus
$-e^{-z} = \frac{t^2}{2} - 1$, or $z = -\ln(1 - \frac{t^2}{2})$.

25. $\frac{dw}{d\theta} = \theta w^2 \sin \theta^2$ implies that $\int \frac{dw}{w^2} = \int \theta \sin \theta^2\, d\theta$ implies that $-\frac{1}{w} = -\frac{1}{2}\cos \theta^2 + C$. According to the initial
conditions, $w(0) = 1$, so $-1 = -\frac{1}{2} + C$ and $C = -\frac{1}{2}$. Thus $-\frac{1}{w} = -\frac{1}{2}\cos \theta^2 - \frac{1}{2}$ implies that $\frac{1}{w} = \frac{\cos \theta^2 + 1}{2}$ implies
that $w = \frac{2}{\cos \theta^2 + 1}$.

Problems

29. $\frac{dQ}{dt} - \frac{Q}{k} = 0$ so $\frac{dQ}{dt} = \frac{Q}{k}$. This is now the same problem as Problem 30, except the constant factor on the right is $\frac{1}{k}$ instead of k. Thus the solution is $Q = Ae^{\frac{1}{k}t}$ for any constant A.

33. Separating variables and integrating gives

$$\int \frac{1}{aP+b} dP = \int dt.$$

This gives

$$\frac{1}{a}\ln|aP+b| = t + C$$
$$\ln|aP+b| = at + D$$
$$aP + b = \pm e^{at+D} = Ae^{at}$$

or

$$P(t) = \frac{1}{a}(Ae^{at} - b).$$

37. Separating variables and integrating gives

$$\int \frac{1}{L-b} dL = \int k(x+a)dx$$

or

$$\ln|L-b| = k\left(\frac{1}{2}x^2 + ax\right) + C.$$

Solving for L gives

$$L(x) = b + Ae^{k\left(\frac{1}{2}x^2 + ax\right)}.$$

41. Since $\frac{dy}{dt} = -y\ln(\frac{y}{2})$, we have $\frac{dy}{y\ln(\frac{y}{2})} = -dt$, so that $\int \frac{dy}{y\ln(\frac{y}{2})} = \int(-dt)$.

Substituting $w = \ln(\frac{y}{2})$, $dw = \frac{1}{y}dy$ gives:

$$\int \frac{dw}{w} = \int(-dt)$$

so

$$\ln|w| = \ln\left|\ln\left(\frac{y}{2}\right)\right| = -t + C.$$

Since $y(0) = 1$, we have $C = \ln|\ln\frac{1}{2}| = \ln|-\ln 2| = \ln(\ln 2)$. Thus $\ln|\ln(\frac{y}{2})| = -t + \ln(\ln 2)$, or

$$\left|\ln\left(\frac{y}{2}\right)\right| = e^{-t+\ln(\ln 2)} = (\ln 2)e^{-t}$$

Again, since $y(0) = 1$, we see that $-\ln(y/2) = (\ln 2)e^{-t}$ and thus $y = 2(2^{-e^{-t}})$. (Note that $\ln(y/2) = (\ln 2)e^{-t}$ does not satisfy $y(0) = 1$.)

45. (a), (b)

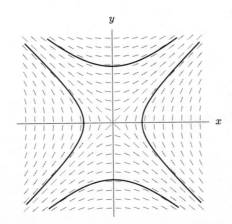

(c) Since $\frac{dy}{dx} = \frac{x}{y}$, we have $\int y\,dy = \int x\,dx$ and thus $\frac{y^2}{2} = \frac{x^2}{2} + C$, or $y^2 - x^2 = 2C$. This is the equation of the hyperbolas in part (b).

Solutions for Section 11.5

Exercises

1. (a) = (I), (b) = (IV), (c) = (III). Graph (II) represents an egg originally at $0°$ C which is moved to the kitchen table ($20°$ C) two minutes after the egg in part (a) is moved.

5. The equilibrium solutions of a differential equation are those functions satisfying the differential equation whose derivative is everywhere 0. Graphically, this means that a function is an equilibrium solution if it is a horizontal line that lies on the slope field. Looking at the figure in the problem, it appears that the equilibrium solutions for this problem are at $y = 1$ and $y = 3$. An equilibrium solution is stable if a small change in the initial value conditions gives a solution which tends toward equilibrium as $t \to \infty$. we see that $y = 3$ is a stable solution, while $y = 1$ is an unstable solution. See Figure 11.4.

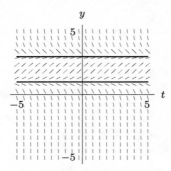

Figure 11.4

Problems

9. (a) Suppose $Y(t)$ is the quantity of oil in the well at time t. We know that the oil in the well decreases at a rate proportional to $Y(t)$, so

$$\frac{dY}{dt} = -kY.$$

Integrating, and using the fact that initially $Y = Y_0 = 10^6$, we have

$$Y = Y_0 e^{-kt} = 10^6 e^{-kt}.$$

In six years, $Y = 500,000 = 5 \cdot 10^5$, so

$$5 \cdot 10^5 = 10^6 e^{-k \cdot 6}$$

so

$$0.5 = e^{-6k}$$
$$k = -\frac{\ln 0.5}{6} = 0.1155.$$

When $Y = 600,000 = 6 \cdot 10^5$,

$$\text{Rate at which oil decreasing} = \left| \frac{dY}{dt} \right| = kY = 0.1155(6 \cdot 10^5) = 69{,}300 \text{ barrels/year.}$$

(b) We solve the equation

$$5 \cdot 10^4 = 10^6 e^{-0.1155t}$$
$$0.05 = e^{-0.1155t}$$
$$t = \frac{\ln 0.05}{-0.1155} = 25.9 \text{ years.}$$

13. (a) $\frac{dB}{dt} = \frac{r}{100}B$. The constant of proportionality is $\frac{r}{100}$.

(b) Solving, we have

$$\frac{dB}{B} = \frac{r\,dt}{100}$$

$$\int \frac{dB}{B} = \int \frac{r}{100}\,dt$$

$$\ln|B| = \frac{r}{100}t + C$$

$$B = e^{(r/100)t+C} = Ae^{(r/100)t}, \qquad A = e^C.$$

A is the initial amount in the account, since A is the amount at time $t = 0$.

(c)

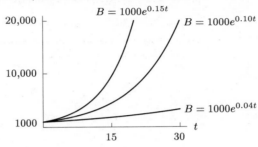

17. (a)

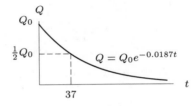

(b) $\dfrac{dQ}{dt} = -kQ$

(c) Since $25\% = 1/4$, it takes two half-lives $= 74$ hours for the drug level to be reduced to 25%. Alternatively, $Q = Q_0e^{-kt}$ and $\frac{1}{2} = e^{-k(37)}$, we have

$$k = -\frac{\ln(1/2)}{37} \approx 0.0187.$$

Therefore $Q = Q_0e^{-0.0187t}$. We know that when the drug level is 25% of the original level that $Q = 0.25Q_0$. Setting these equal, we get

$$0.25 = e^{-0.0187t}.$$

giving

$$t = -\frac{\ln(0.25)}{0.0187} \approx 74 \text{ hours} \approx 3 \text{ days}.$$

21. The rate of disintegration is proportional to the quantity of carbon-14 present. Let Q be the quantity of carbon-14 present at time t, with $t = 0$ in 1977. Then

$$Q = Q_0e^{-kt},$$

where Q_0 is the quantity of carbon-14 present in 1977 when $t = 0$. Then we know that

$$\frac{Q_0}{2} = Q_0e^{-k(5730)}$$

so that

$$k = -\frac{\ln(1/2)}{5730} = 0.000121.$$

Thus

$$Q = Q_0e^{-0.000121t}.$$

The quantity present at any time is proportional to the rate of disintegration at that time so

$$Q_0 = c8.2 \quad \text{and} \quad Q = c13.5$$

where c is a constant of proportionality. Thus substituting for Q and Q_0 in

$$Q = Q_0 e^{-0.000121t}$$

gives

$$c13.5 = c8.2e^{-0.000121t}$$

so

$$t = -\frac{\ln(13.5/8.2)}{0.000121} \approx -4120.$$

Thus Stonehenge was built about 4120 years before 1977, in about 2150 B.C.

Solutions for Section 11.6

Exercises

1. Since mg is constant and $a = dv/dt$, differentiating $ma = mg - kv$ gives

$$m\frac{da}{dt} = -k\frac{dv}{dt} = -ma.$$

Thus, the differential equation is

$$\frac{da}{dt} = -\frac{k}{m}a.$$

Solving for a gives

$$a = a_0 e^{-kt/m}.$$

At $t = 0$, we have $a = g$, the acceleration due to gravity. Thus, $a_0 = g$, so

$$a = ge^{-kt/m}.$$

Problems

5. Let $D(t)$ be the quantity of dead leaves, in grams per square centimeter. Then $\frac{dD}{dt} = 3 - 0.75D$, where t is in years. We factor out -0.75 and then separate variables.

$$\frac{dD}{dt} = -0.75(D - 4)$$

$$\int \frac{dD}{D - 4} = \int -0.75 \, dt$$

$$\ln |D - 4| = -0.75t + C$$

$$|D - 4| = e^{-0.75t+C} = e^{-0.75t}e^C$$

$$D = 4 + Ae^{-0.75t}, \quad \text{where } A = \pm e^C.$$

If initially the ground is clear, the solution looks like the following graph:

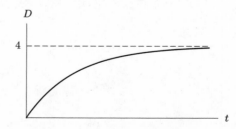

The equilibrium level is 4 grams per square centimeter, regardless of the initial condition.

9. We are given that
$$BC = 2OC.$$

If the point A has coordinates (x, y) then $OC = x$ and $AC = y$. The slope of the tangent line, y', is given by
$$y' = \frac{AC}{BC} = \frac{y}{BC},$$
so
$$BC = \frac{y}{y'}.$$

Substitution into $BC = 2OC$ gives
$$\frac{y}{y'} = 2x,$$
so
$$\frac{y'}{y} = \frac{1}{2x}.$$

Separating variables to integrate this differential equation gives
$$\int \frac{dy}{y} = \int \frac{dx}{2x}$$
$$\ln|y| = \frac{1}{2}\ln|x| + C = \ln\sqrt{|x|} + \ln A$$
$$|y| = A\sqrt{|x|}$$
$$y = \pm(A\sqrt{x}).$$

Thus, in the first quadrant, the curve has equation $y = A\sqrt{x}$.

13. (a) If I is intensity and l is the distance traveled through the water, then for some $k > 0$,
$$\frac{dI}{dl} = -kI.$$

(The proportionality constant is negative because intensity decreases with distance). Thus $I = Ae^{-kl}$. Since $I = A$ when $l = 0$, A represents the initial intensity of the light.

(b) If 50% of the light is absorbed in 10 feet, then $0.50A = Ae^{-10k}$, so $e^{-10k} = \frac{1}{2}$, giving
$$k = \frac{-\ln\frac{1}{2}}{10} = \frac{\ln 2}{10}.$$

In 20 feet, the percentage of light left is
$$e^{-\frac{\ln 2}{10} \cdot 20} = e^{-2\ln 2} = (e^{\ln 2})^{-2} = 2^{-2} = \frac{1}{4},$$

so $\frac{3}{4}$ or 75% of the light has been absorbed. Similarly, after 25 feet,
$$e^{-\frac{\ln 2}{10} \cdot 25} = e^{-2.5\ln 2} = (e^{\ln 2})^{-\frac{5}{2}} = 2^{-\frac{5}{2}} \approx 0.177.$$

Approximately 17.7% of the light is left, so 82.3% of the light has been absorbed.

17. (a) The quantity and the concentration both increase with time. As the concentration increases, the rate at which the drug is excreted also increases, and so the rate at which the drug builds up in the blood decreases; thus the graph of concentration against time is concave down. The concentration rises until the rate of excretion exactly balances the rate at which the drug is entering; at this concentration there is a horizontal asymptote. (See Figure 11.5.)

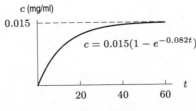

Figure 11.5

(b) Let's start by writing a differential equation for the quantity, $Q(t)$.

$$\text{Rate at which quantity of drug changes} = \text{Rate in} - \text{Rate out}$$

$$\frac{dQ}{dt} = 43.2 - 0.082Q$$

where Q is measured in mg. We want an equation for concentration $c(t) = Q(t)/v$, where $c(t)$ is measured in mg/ml and v is volume, so $v = 35,000$ ml.

$$\frac{1}{v}\frac{dQ}{dt} = \frac{43.2}{v} - 0.082\frac{Q}{v},$$

giving

$$\frac{dc}{dt} = \frac{43.2}{35,000} - 0.082c.$$

(c) Factor out -0.082 and separate variables to solve.

$$\frac{dc}{dt} = -0.082(c - 0.015)$$

$$\int \frac{dc}{c - 0.015} = -0.082 \int dt$$

$$\ln|c - 0.015| = -0.082t + B$$

$$c - 0.015 = Ae^{-0.082t} \quad \text{where} \quad A = \pm e^B$$

Since $c = 0$ when $t = 0$, we have $A = -0.015$, so

$$c = 0.015 - 0.015e^{-0.082t} = 0.015(1 - e^{-0.082t}).$$

Thus $c \to 0.015$ mg/ml as $t \to \infty$.

21. (a) Concentration of carbon monoxide $= \dfrac{\text{Quantity in room}}{\text{Volume}}$.

If $Q(t)$ represents the quantity of carbon monoxide in the room at time t, $c(t) = Q(t)/60$.

$$\begin{array}{c}\text{Rate quantity of}\\ \text{carbon monoxide in room} \\ \text{changes}\end{array} = \text{rate in} - \text{rate out}$$

Now

$$\text{Rate in} = 5\%(0.002\text{m}^3/\text{min}) = 0.05(0.002) = 0.0001\text{m}^3/\text{min}.$$

Since smoky air is leaving at $0.002\text{m}^3/\text{min}$, containing a concentration $c(t) = Q(t)/60$ of carbon monoxide

$$\text{Rate out} = 0.002\frac{Q(t)}{60}$$

Thus

$$\frac{dQ}{dt} = 0.0001 - \frac{0.002}{60}Q$$

Since $c = Q/60$, we can substitute $Q = 60c$, giving

$$\frac{d(60c)}{dt} = 0.0001 - \frac{0.002}{60}(60c)$$

$$\frac{dc}{dt} = \frac{0.0001}{60} - \frac{0.002}{60}c$$

(b) Factoring the right side of the differential equation and separating gives

$$\frac{dc}{dt} = -\frac{0.0001}{3}(c - 0.05) \approx 3 \times 10^{-5}(c - 0.05)$$

$$\int \frac{dc}{c - 0.05} = -\int 3 \times 10^{-5}dt$$

$$\ln|c - 0.05| = -3 \times 10^{-5}t + K$$

$$c - 0.05 = Ae^{-3\times10^{-5}t} \quad \text{where} A = \pm e^K.$$

Since $c = 0$ when $t = 0$, we have $A = -0.05$, so

$$c = 0.05 - 0.05e^{-3\times10^{-5}t}$$

(c) As $t \to \infty$, $e^{-3\times10^{-5}t} \to 0$ so $c \to 0.05$.

Thus in the long run, the concentration of carbon monoxide tends to 5%, the concentration of the incoming air.

Solutions for Section 11.7

Exercises

1. A continuous growth rate of 0.2% means that

$$\frac{1}{P}\frac{dP}{dt} = 0.2\% = 0.002.$$

Separating variables and integrating gives

$$\int \frac{dP}{P} = \int 0.002\, dt$$

$$P = P_0 e^{0.002t} = (6.6 \times 10^6)e^{0.002t}.$$

Problems

5.

Table 11.7

Year	P	$\frac{dP}{dt} \approx \frac{P(t+10)-P(t-10)}{20}$
1790	3.9	
1800	5.3	$(7.2 - 3.9)/20 = 0.165$
1810	7.2	$(9.6 - 5.3)/20 = 0.215$
1820	9.6	$(12.9 - 7.2)/20 = 0.285$
1830	12.9	$(17.1 - 9.6)/20 = 0.375$
1840	17.1	$(23.2 - 12.9)/20 = 0.515$
1850	23.2	$(31.4 - 17.1)/20 = 0.715$
1860	31.4	$(38.6 - 23.2)/20 = 0.770$
1870	38.6	$(50.2 - 31.4)/20 = 0.940$
1880	50.2	$(62.9 - 38.6)/20 = 1.215$
1890	62.9	$(76.0 - 50.2)/20 = 1.290$
1900	76.0	$(92.0 - 62.9)/20 = 1.455$
1910	92.0	$(105.7 - 76.0)/20 = 1.485$
1920	105.7	$(122.8 - 92.0)/20 = 1.540$
1930	122.8	$(131.7 - 105.7)/20 = 1.300$
1940	131.7	$(150.7 - 122.8)/20 = 1.395$
1950	150.7	

According to these calculations, the largest value of dP/dt occurs in 1920 when the rate of change is $\frac{dP}{dt} = 1.540$ million people/year. The population in 1920 was 105.7 million. If we assume that the limiting value, L, is twice the population when it is changing most quickly, then $L = 2 \times 105.7 = 211.4$ million. This is greater than the estimate of 187 million computed in the text and closer to the actual 1990 population of 248.7 million.

9. **(a)** Let I be the number of informed people at time t, and I_0 the number who know initially. Then this model predicts that $\frac{dI}{dt} = k(M - I)$ for some positive constant k. Solving this, we find the solution is

$$I = M - (M - I_0)e^{-kt}.$$

We sketch the solution with $I_0 = 0$. Notice that $\frac{dI}{dt}$ is largest when I is smallest, so the information spreads fastest in the beginning, at $t = 0$. In addition, the graph below shows that $I \to M$ as $t \to \infty$, meaning that everyone gets the information eventually.

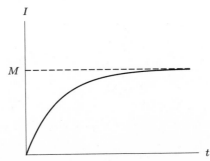

(b) In this case, the model suggests that $\frac{dI}{dt} = kI(M - I)$ for some positive constant k. This is a logistic model with carrying capacity M. We sketch the solutions for three different values of I_0 below.

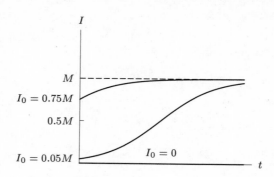

(i) If $I_0 = 0$ then $I = 0$ for all t. In other words, if nobody knows something, it doesn't spread by word of mouth!

(ii) If $I_0 = 0.05M$, then $\frac{dI}{dt}$ is increasing up to $I = \frac{M}{2}$. Thus, the information is spreading fastest at $I = \frac{M}{2}$.

(iii) If $I_0 = 0.75M$, then $\frac{dI}{dt}$ is always decreasing for $I > \frac{M}{2}$, so $\frac{dI}{dt}$ is largest when $t = 0$.

13.

(a)

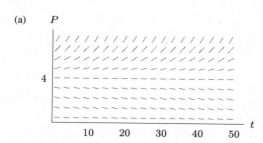

(b)

(c) There are two equilibrium values, $P = 0$, and $P = 4$. The first, representing extinction, is stable. The equilibrium value $P = 4$ is unstable because the populations increase if greater than 4, and decrease if less than 4. Notice that the equilibrium values can be obtained by setting $dP/dt = 0$:

$$\frac{dP}{dt} = 0.02P^2 - 0.08P = 0.02P(P - 4) = 0$$

so

$$P = 0 \text{ or } P = 4.$$

Solutions for Section 11.8

Exercises

1. Since

$$\frac{dS}{dt} = -aSI,$$

$$\frac{dI}{dt} = aSI - bI,$$

$$\frac{dR}{dt} = bI$$

we have

$$\frac{dS}{dt} + \frac{dI}{dt} + \frac{dR}{dt} = -aSI + aSI - bI + bI = 0.$$

Thus $\frac{d}{dt}(S + I + R) = 0$, so $S + I + R = $ constant.

5. If $w = 2$ and $r = 2$, then $\frac{dw}{dt} = -2$ and $\frac{dr}{dt} = 2$, so initially the number of worms decreases and the number of robins increases. In the long run, however, the populations will oscillate; they will even go back to $w = 2$ and $r = 2$.

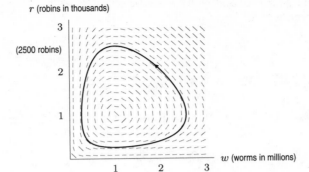

9. The numbers of robins begins to increase while the number of worms remains approximately constant. See Figure 11.6. The numbers of robins and worms oscillate periodically between 0.2 and 3, with the robin population lagging behind the worm population.

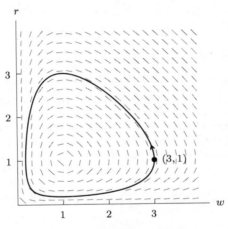

Figure 11.6

13. x decreases quickly while y increases more slowly.

Problems

17. (a) Predator-prey, because x decreases while alone, but is helped by y, whereas y increases logistically when alone, and is harmed by x. Thus x is predator, y is prey.

(b)

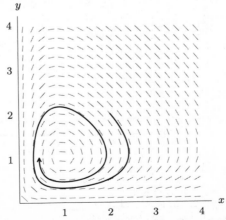

Provided neither initial population is zero, both populations tend to about 1. If x is initially zero, but y is not, then $y \to \infty$. If y is initially zero, but x is not, then $x \to 0$.

21. (a) Since the guerrillas are hard to find, the rate at which they are put out of action is proportional to the number of chance encounters between a guerrilla and a conventional soldier, which is in turn proportional to the number of guerrillas and to the number of conventional soldiers. Thus the rate at which guerrillas are put out of action is proportional to the product of the strengths of the two armies.

(b)

$$\frac{dx}{dt} = -xy$$

$$\frac{dy}{dt} = -x$$

(c) Thinking of y as a function of x and x a function of of t, then by the chain rule: $\frac{dy}{dt} = \frac{dy}{dx}\frac{dx}{dt}$ so:

$$\frac{dy}{dx} = \frac{dy/dt}{dx/dt} = \frac{-x}{-xy} = \frac{1}{y}$$

Separating variables:

$$\int y \, dy = \int dx$$

$$\frac{y^2}{2} = x + C$$

The value of C is determined by the initial strengths of the two armies.

(d) The sign of C determines which side wins the battle. Looking at the general solution $\frac{y^2}{2} = x + C$, we see that if $C > 0$ the y-intercept is at $\sqrt{2C}$, so y wins the battle by virtue of the fact that it still has troops when $x = 0$. If $C < 0$ then the curve intersects the axes at $x = -C$, so x wins the battle because it has troops when $y = 0$. If $C = 0$, then the solution goes to the point $(0, 0)$, which represents the case of mutual annihilation.

(e) We assume that an army wins if the opposing force goes to 0 first. Figure 11.7 shows that the conventional force wins if $C > 0$ and the guerrillas win if $C < 0$. Neither side wins if $C = 0$ (all soldiers on both sides are killed in this case).

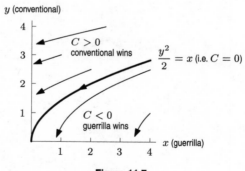

Figure 11.7

Solutions for Section 11.9

Exercises

1. (a) $dS/dt = 0$ where $S = 0$ or $I = 0$ (both axes).

$dI/dt = 0.0026I(S - 192)$, so $dI/dt = 0$ where $I = 0$ or $S = 192$.

Thus every point on the S axis is an equilibrium point (corresponding to no one being sick).

(b) In region I, where $S > 192$, $\dfrac{dS}{dt} < 0$ and $\dfrac{dI}{dt} > 0$.

In region II, where $S < 192$, $\dfrac{dS}{dt} < 0$ and $\dfrac{dI}{dt} < 0$. See Figure 11.8.

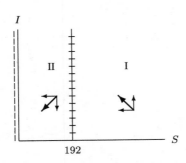

Figure 11.8

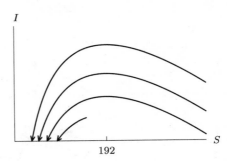

Figure 11.9

(c) If the trajectory starts with $S_0 > 192$, then I increases to a maximum when $S = 192$. If $S_0 < 192$, then I always decreases. See Figure 11.8. Regardless of the initial conditions, the trajectory always goes to a point on the S-axis (where $I = 0$). The S-intercept represents the number of students who never get the disease. See Figure 11.9.

Problems

5. We first find the nullclines. Vertical nullclines occur where $\frac{dx}{dt} = 0$, which happens when $x = 0$ or $y = \frac{1}{3}(2 - x)$. Horizontal nullclines occur where $\frac{dy}{dt} = y(1 - 2x) = 0$, which happens when $y = 0$ or $x = \frac{1}{2}$. These nullclines are shown in Figure 11.10.

Equilibrium points (also shown in Figure 11.10) occur at the intersections of vertical and horizontal nullclines. There are three such points for this system of equations; $(0, 0)$, $(\frac{1}{2}, \frac{1}{2})$ and $(2, 0)$.

The nullclines divide the positive quadrant into four regions as shown in Figure 11.10. Trajectory directions for these regions are shown in Figure 11.11.

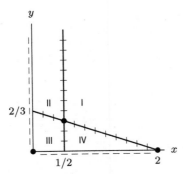

Figure 11.10: Nullclines and equilibrium points (dots)

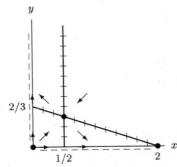

Figure 11.11: General directions of trajectories and equilibrium points (dots)

9. We assume that x, $y \geq 0$ and then find the nullclines. $\frac{dx}{dt} = x(1 - \frac{x}{2} - y) = 0$ when $x = 0$ or $y + \frac{x}{2} = 1$.
$\frac{dy}{dt} = y(1 - \frac{y}{3} - x) = 0$ when $y = 0$ or $x + \frac{y}{3} = 1$.
We find the equilibrium points. They are $(2, 0)$, $(0, 3)$, $(0, 0)$, and $(\frac{4}{5}, \frac{3}{5})$. The nullclines and equilibrium points are shown in Figure 11.12.

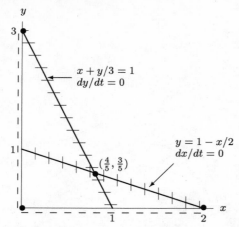

Figure 11.12: Nullclines and equilibrium points (dots)

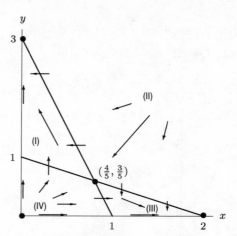

Figure 11.13: General directions of trajectories and equilibrium points (dots)

Figure 11.13 shows that if the initial point is in sector (I), the trajectory heads towards the equilibrium point $(0, 3)$. Similarly, if the trajectory begins in sector (III), then it heads towards the equilibrium $(2, 0)$ over time. If the trajectory begins in sector (II) or (IV), it can go to any of the three equilibrium points $(2, 0)$, $(0, 3)$, or $\left(\frac{4}{5}, \frac{3}{5}\right)$.

13. (a)

$$\frac{dx}{dt} = 0 \text{ when } x = \frac{10.5}{0.45} = 23.3$$

$$\frac{dy}{dt} = 0 \text{ when } 8.2x - 0.8y - 142 = 0$$

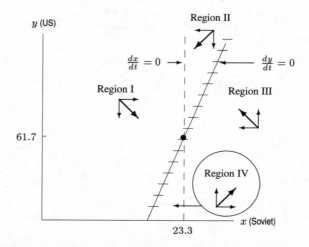

Figure 11.14: Nullclines and equilibrium point (dot) for US-Soviet arms race

There is an equilibrium point where the trajectories cross at $x = 23.3$, $y = 61.7$

In region I, $\dfrac{dx}{dt} > 0$, $\dfrac{dy}{dt} < 0$.

In region II, $\dfrac{dx}{dt} < 0$, $\dfrac{dy}{dt} < 0$.

In region III, $\dfrac{dx}{dt} < 0$, $\dfrac{dy}{dt} > 0$.

In region IV, $\dfrac{dx}{dt} > 0$, $\dfrac{dy}{dt} > 0$.

(b)

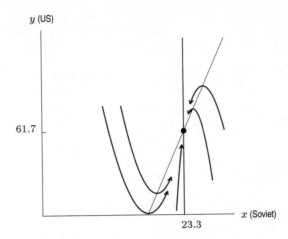

y (US)

61.7

23.3

x (Soviet)

Figure 11.15: Trajectories for US-Soviet arms race.

(c) All the trajectories tend towards the equilibrium point $x = 23.3$, $y = 61.7$. Thus the model predicts that in the long run the arms race will level off with the Soviet Union spending 23.3 billion dollars a year on arms and the US 61.7 billion dollars.

(d) As the model predicts, yearly arms expenditure did tend towards 23 billion for the Soviet Union and 62 billion for the US.

Solutions for Section 11.10

Exercises

1. If $y = 2\cos t + 3\sin t$, then $y' = -2\sin t + 3\cos t$ and $y'' = -2\cos t - 3\sin t$. Thus, $y'' + y = 0$.

5. If $y(t) = A\sin(\omega t) + B\cos(\omega t)$ then
$$y' = \omega A\cos(\omega t) - \omega B\sin(\omega t)$$
$$y'' = -\omega^2 A\sin(\omega t) - \omega^2 B\cos(\omega t)$$

therefore
$$y'' + \omega^2 y = -\omega^2 A\sin(\omega t) - \omega^2 B\cos(2t) + \omega^2(A\sin(\omega t) + B\cos(\omega t)) = 0$$

for all values of A and B, so the given function is a solution.

9. The amplitude is $\sqrt{3^2 + 7^2} = \sqrt{58}$.

Problems

13. At $t = 0$, we find that $y = 2$, which is clearly the highest point since $-1 \le \cos 3t \le 1$. Thus, at $t = 0$ the mass is at its highest point. Since $y' = -6\sin 3t$, we see $y' = 0$ when $t = 0$. Thus, at $t = 0$ the object is at rest, although it will move down after $t = 0$.

17. First, we note that the solutions of:
(a) $x'' + x = 0$ are $x = A\cos t + B\sin t$;
(b) $x'' + 4x = 0$ are $x = A\cos 2t + B\sin 2t$;
(c) $x'' + 16x = 0$ are $x = A\cos 4t + B\sin 4t$.
This follows from what we know about the general solution to $x'' + \omega^2 x = 0$.
The period of the solutions to (a) is 2π, the period of the solutions to (b) is π, and the period of the solutions of (c) is $\frac{\pi}{2}$. Since the t-scales are the same on all of the graphs, we see that graphs (I) and (IV) have the same period, which is twice the period of graph (III). Graph (II) has twice the period of graphs (I) and (IV). Since each graph represents a solution, we have the following:

- equation (a) goes with graph (II)
  equation (b) goes with graphs (I) and (IV)
  equation (c) goes with graph (III)

- The graph of (I) passes through $(0, 0)$, so $0 = A \cos 0 + B \sin 0 = A$. Thus, the equation is $x = B \sin 2t$. Since the amplitude is 2, we see that $x = 2 \sin 2t$ is the equation of the graph. Similarly, the equation for (IV) is $x = -3 \sin 2t$. The graph of (II) also passes through $(0, 0)$, so, similarly, the equation must be $x = B \sin t$. In this case, we see that $B = -1$, so $x = -\sin t$.

  Finally, the graph of (III) passes through $(0, 1)$, and 1 is the maximum value. Thus, $1 = A \cos 0 + B \sin 0$, so $A = 1$. Since it reaches a local maximum at $(0, 1)$, $x'(0) = 0 = -4A \sin 0 + 4B \cos 0$, so $B = 0$. Thus, the solution is $x = \cos 4t$.

21. (a) Since a mass of 3 kg stretches the spring by 2 cm, the spring constant k is given by

$$3g = 2k \quad \text{so} \quad k = \frac{3g}{2}.$$

See Figure 11.16.

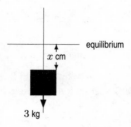

Figure 11.16

Suppose we measure the displacement x from the equilibrium; then, using

$$\text{Mass} \cdot \text{Acceleration} = \text{Force}$$

gives

$$3x'' = -kx = -\frac{3gx}{2}$$

$$x'' + \frac{g}{2}x = 0$$

Since at time $t = 0$, the brick is 5 cm below the equilibrium and not moving, the initial conditions are $x(0) = 5$ and $x'(0) = 0$.

(b) The solution to the differential equation is

$$x = A \cos\left(\sqrt{\frac{g}{2}}t\right) + B \sin\left(\sqrt{\frac{g}{2}}t\right).$$

Since $x(0) = 5$, we have

$$x = A \cos(0) + B \sin(0) = 5 \quad \text{so} \quad A = 5.$$

In addition,

$$x'(t) = -5\sqrt{\frac{g}{2}} \sin\left(\sqrt{\frac{g}{2}}t\right) + B\sqrt{\frac{g}{2}} \cos\left(\sqrt{\frac{g}{2}}t\right)$$

so

$$x'(0) = -5\sqrt{\frac{g}{2}} \sin(0) + B\sqrt{\frac{g}{2}} \cos(0) = 0 \quad \text{so} \quad B = 0.$$

Thus,

$$x = 5 \cos \sqrt{\frac{g}{2}}t.$$

25. The equation we have for the charge tells us that:

$$\frac{d^2Q}{dt^2} = -\frac{Q}{LC},$$

where L and C are positive.

If we let $\omega = \sqrt{\frac{1}{LC}}$, we know the solution is of the form:

$$Q = C_1 \cos \omega t + C_2 \sin \omega t.$$

Since $Q(0) = 0$, we find that $C_1 = 0$, so $Q = C_2 \sin \omega t$.

Since $Q'(0) = 4$, and $Q' = \omega C_2 \cos \omega t$, we have $C_2 = \dfrac{4}{\omega}$, so $Q = \dfrac{4}{\omega} \sin \omega t$.

But we want the maximum charge, meaning the amplitude of Q, to be $2\sqrt{2}$ coulombs. Thus, we have $\dfrac{4}{\omega} = 2\sqrt{2}$, which

gives us $\omega = \sqrt{2}$.

So we now have: $\sqrt{2} = \dfrac{1}{\sqrt{LC}} = \dfrac{1}{\sqrt{10C}}$. Thus, $C = \frac{1}{20}$ farads.

Solutions for Section 11.11

Exercises

1. The characteristic equation is $r^2 + 4r + 3 = 0$, so $r = -1$ or -3.
   Therefore $y(t) = C_1 e^{-t} + C_2 e^{-3t}$.

5. The characteristic equation is $r^2 + 7 = 0$, so $r = \pm\sqrt{7}i$.
   Therefore $s(t) = C_1 \cos \sqrt{7}t + C_2 \sin \sqrt{7}t$.

9. The characteristic equation is $r^2 + r + 1 = 0$, so $r = -\frac{1}{2} \pm \frac{\sqrt{3}}{2}i$.
   Therefore $p(t) = C_1 e^{-t/2} \cos \frac{\sqrt{3}}{2}t + C_2 e^{-t/2} \sin \frac{\sqrt{3}}{2}t$.

13. The characteristic equation is
$$r^2 + 5r + 6 = 0$$
which has the solutions $r = -2$ and $r = -3$ so that
$$y(t) = Ae^{-3t} + Be^{-2t}$$

The initial condition $y(0) = 1$ gives
$$A + B = 1$$

and $y'(0) = 0$ gives
$$-3A - 2B = 0$$

so that $A = -2$ and $B = 3$ and
$$y(t) = -2e^{-3t} + 3e^{-2t}$$

17. The characteristic equation is $r^2 + 6r + 5 = 0$, so $r = -1$ or -5.
    Therefore $y(t) = C_1 e^{-t} + C_2 e^{-5t}$.
    $y'(t) = -C_1 e^{-t} - 5C_2 e^{-5t}$
    $y'(0) = 0 = -C_1 - 5C_2$
    $y(0) = 1 = C_1 + C_2$
         Therefore $C_2 = -1/4$, $C_1 = 5/4$ and $y(t) = \frac{5}{4}e^{-t} - \frac{1}{4}e^{-5t}$.

21. The characteristic equation is
$$r^2 + 5r + 6 = 0$$
which has the solutions $r = -2$ and $r = -3$ so that
$$y(t) = Ae^{-2t} + Be^{-3t}$$

The initial condition $y(0) = 1$ gives
$$A + B = 1$$

and $y(1) = 0$ gives
$$Ae^{-2} + Be^{-3} = 0$$

so that $A = \dfrac{1}{1 - e}$ and $B = -\dfrac{e}{1 - e}$ and

$$y(t) = \frac{1}{1 - e}e^{-2t} + \frac{-e}{1 - e}e^{-3t}$$

Problems

25. (a) $x'' + 4x = 0$ represents an undamped oscillator, and so goes with (IV).

(b) $x'' - 4x = 0$ has characteristic equation $r^2 - 4 = 0$ and so $r = \pm 2$. The solution is $C_1 e^{-2t} + C_2 e^{2t}$. This represents non-oscillating motion, so it goes with (II).

(c) $x'' - 0.2x' + 1.01x = 0$ has characteristic equation $r^2 - 0.2 + 1.01 = 0$ so $b^2 - 4ac = 0.04 - 4.04 = -4$, and $r = 0.1 \pm i$. So the solution is

$$C_1 e^{(0.1+i)t} + C_2 e^{(0.1-i)t} = e^{0.1t}(A \sin t + B \cos t).$$

The negative coefficient in the x' term represents an amplifying force. This is reflected in the solution by $e^{0.1t}$, which increases as t increases, so this goes with (I).

(d) $x'' + 0.2x' + 1.01x$ has characteristic equation $r^2 + 0.2r + 1.01 = 0$ so $b^2 - 4ac = -4$. This represents a damped oscillator. We have $r = -0.1 \pm i$ and so the solution is $x = e^{-0.1t}(A \sin t + B \cos t)$, which goes with (III).

29. Recall that $F_{\text{drag}} = -c\frac{ds}{dt}$, so to find the largest coefficient of damping we look at the coefficient of s'. Thus spring (iii) has the largest coefficient of damping.

33. The stiffest spring exerts the greatest restoring force for a small displacement. Recall that by Hooke's Law $F_{\text{spring}} = -ks$, so we look for the differential equation with the greatest coefficient of s. This is spring (ii).

37. The characteristic equation is $r^2 + r - 2 = 0$, so $r = 1$ or -2. Therefore $z(t) = C_1 e^t + C_2 e^{-2t}$. Since $e^t \to \infty$ as $t \to \infty$, we must have $C_1 = 0$. Therefore $z(t) = C_2 e^{-2t}$. Furthermore, $z(0) = 3 = C_2$, so $z(t) = 3e^{-2t}$.

41. In this case, the differential equation describing charge is $8Q'' + 2Q' + \frac{1}{4}Q = 0$, so the characteristic equation is $8r^2 + 2r + \frac{1}{4} = 0$. This quadratic equation has solutions

$$r = \frac{-2 \pm \sqrt{4 - 4 \cdot 8 \cdot \frac{1}{4}}}{16} = -\frac{1}{8} \pm \frac{1}{8}i.$$

Thus, the equation for charge is

$$Q(t) = e^{-\frac{1}{8}t}\left(A \sin \frac{t}{8} + B \cos \frac{t}{8}\right).$$

$$Q'(t) = -\frac{1}{8}e^{-\frac{1}{8}t}\left(A \sin \frac{t}{8} + B \cos \frac{t}{8}\right) + e^{-\frac{1}{8}t}\left(\frac{1}{8}A \cos \frac{t}{8} - \frac{1}{8}B \sin \frac{t}{8}\right)$$

$$= \frac{1}{8}e^{-\frac{1}{8}t}\left((A - B) \cos \frac{t}{8} + (-A - B) \sin \frac{t}{8}\right).$$

(a) We have

$$Q(0) = B = 0,$$

$$Q'(0) = \frac{1}{8}(A - B) = 2.$$

Thus, $B = 0$, $A = 16$, and

$$Q(t) = 16e^{-\frac{1}{8}t} \sin \frac{t}{8}.$$

(b) We have

$$Q(0) = B = 2,$$

$$Q'(0) = \frac{1}{8}(A - B) = 0.$$

Thus, $B = 2$, $A = 2$, and

$$Q(t) = 2e^{-\frac{1}{8}t}\left(\sin \frac{t}{8} + \cos \frac{t}{8}\right).$$

(c) By increasing the inductance, we have gone from the overdamped case to the underdamped case. We find that while the charge still tends to 0 as $t \to \infty$, the charge in the underdamped case oscillates between positive and negative values. In the over-damped case of Problem 39, the charge starts nonnegative and remains positive.

Solutions for Chapter 11 Review

Exercises

1. (a) Yes   **(b)** No   **(c)** Yes
    (d) No   **(e)** Yes   **(f)** Yes
    (g) No   **(h)** Yes   **(i)** No
    (j) Yes   **(k)** Yes   **(l)** No

5. This equation is separable, so we integrate, giving

$$\int \frac{1}{10 + 0.5H} \, dH = \int dt$$

so

$$\frac{1}{0.5} \ln |10 + 0.5H| = t + C.$$

Thus

$$H = Ae^{0.5t} - 20.$$

9. $\frac{dP}{dt} = 0.03P + 400$ so $\int \frac{dP}{P + \frac{40000}{3}} = \int 0.03 dt$.
$\ln \left| P + \frac{40000}{3} \right| = 0.03t + C$ giving $P = Ae^{0.03t} - \frac{40000}{3}$. Since $P(0) = 0$, $A = \frac{40000}{3}$, therefore $P = \frac{40000}{3}(e^{0.03t} - 1)$.

13. $\frac{dy}{dx} = \frac{y(3-x)}{x(\frac{1}{2}y - 4)}$ gives $\int \frac{(\frac{1}{2}y - 4)}{y} dy = \int \frac{(3-x)}{x} dx$ so $\int (\frac{1}{2} - \frac{4}{y}) dy = \int (\frac{3}{x} - 1) dx$. Thus $\frac{1}{2}y - 4 \ln |y| = 3 \ln |x| - x + C$.
Since $y(1) = 5$, we have $\frac{5}{2} - 4 \ln 5 = \ln |1| - 1 + C$ so $C = \frac{7}{2} - 4 \ln 5$. Thus,

$$\frac{1}{2}y - 4 \ln |y| = 3 \ln |x| - x + \frac{7}{2} - 4 \ln 5.$$

We cannot solve for y in terms of x, so we leave the equation in this form.

17. $\frac{dy}{dx} = \frac{y(100-x)}{x(20-y)}$ gives $\int (\frac{20-y}{y}) dy = \int (\frac{100-x}{x}) dx$. Thus, $20 \ln |y| - y = 100 \ln |x| - x + C$. The curve passes through $(1, 20)$, so $20 \ln 20 - 20 = -1 + C$ giving $C = 20 \ln 20 - 19$. Therefore, $20 \ln |y| - y = 100 \ln |x| - x + 20 \ln 20 - 19$.
We cannot solve for y in terms of x, so we leave the equation in this form.

21. $e^{-\cos \theta} \frac{dz}{d\theta} = \sqrt{1 - z^2} \sin \theta$ implies $\int \frac{dz}{\sqrt{1-z^2}} = \int e^{\cos \theta} \sin \theta \, d\theta$ implies $\arcsin z = -e^{\cos \theta} + C$. According to the
initial conditions: $z(0) = \frac{1}{2}$, so $\arcsin \frac{1}{2} = -e^{\cos 0} + C$, therefore $\frac{\pi}{6} = -e + C$, and $C = \frac{\pi}{6} + e$. Thus $z = \sin(-e^{\cos \theta} + \frac{\pi}{6} + e)$.

25. The characteristic equation of $9z'' - z = 0$ is

$$9r^2 - 1 = 0.$$

If this is written in the form $r^2 + br + c = 0$, we have that $r^2 - 1/9 = 0$ and

$$b^2 - 4c = 0 - (4)(-1/9) = 4/9 > 0$$

This indicates overdamped motion and since the roots of the characteristic equation are $r = \pm 1/3$, the general solution is

$$y(t) = C_1 e^{\frac{1}{3}t} + C_2 e^{-\frac{1}{3}t}.$$

29. The characteristic equation of $x'' + 2x' + 10x = 0$ is

$$r^2 + 2r + 10 = 0$$

We have that

$$b^2 - 4c = 2^2 - 4(10) = -36 < 0$$

This indicates underdamped motion and since the roots of the characteristic equation are $r = -1 \pm 3i$, the general solution is

$$y(t) = C_1 e^{-t} \cos 3t + C_2 e^{-t} \sin 3t$$

Problems

33. Recall that $s'' + bs' + cs = 0$ is overdamped if the discriminant $b^2 - 4c > 0$, critically damped if $b^2 - 4c = 0$, and underdamped if $b^2 - 4c < 0$. This has discriminant $b^2 - 4c = b^2 + 64$. Since $b^2 + 64$ is always positive, the solution is always overdamped.

37. Let $V(t)$ be the volume of water in the tank at time t, then

$$\frac{dV}{dt} = k\sqrt{V}$$

This is a separable equation which has the solution

$$V(t) = (\frac{kt}{2} + C)^2$$

Since $V(0) = 200$ this gives $200 = C^2$ so

$$V(t) = (\frac{kt}{2} + \sqrt{200})^2.$$

However, $V(1) = 180$ therefore

$$180 = (\frac{k}{2} + \sqrt{200})^2,$$

so that $k = 2\left(\sqrt{180} - \sqrt{200}\right) = -1.45146$. Therefore,

$$V(t) = (-0.726t + \sqrt{200})^2.$$

The tank will be half-empty when $V(t) = 100$, so we solve

$$100 = (-0.726t + \sqrt{200})^2$$

to obtain $t = 5.7$ days. The tank will be half empty in 5.7 days.
The volume after 4 days is $V(4)$ which is approximately 126.32 liters.

41. Let I be the number of infected people. Then, the number of healthy people in the population is $M - I$. The rate of infection is

$$\text{Infection rate} = \frac{0.01}{M}(M - I)I.$$

and the rate of recovery is

$$\text{Recovery rate} = 0.009I.$$

Therefore,

$$\frac{dI}{dt} = \frac{0.01}{M}(M - I)I - 0.009I$$

or

$$\frac{dI}{dt} = 0.001I(1 - 10\frac{I}{M}).$$

This is a logistic differential equation, and so the solution will look like the following graph:

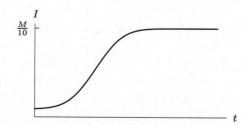

The limiting value for I is $\frac{1}{10}M$, so 1/10 of the population is infected in the long run.

CAS Challenge Problems

45. **(a)** We find the equilibrium solutions by setting $dP/dt = 0$, that is, $P(P - 1)(2 - P) = 0$, which gives three solutions, $P = 0$, $P = 1$, and $P = 2$.

(b) To get your computer algebra system to check that P_1 and P_2 are solutions, substitute one of them into the equation and form an expression consisting of the difference between the right and left hand sides, then ask the CAS to simplify that expression. Do the same for the other function. In order to avoid too much typing, define P_1 and P_2 as functions in your system.

(c) Substituting $t = 0$ gives

$$P_1(0) = 1 - \frac{1}{\sqrt{4}} = 1/2$$

$$P_2(0) = 1 + \frac{1}{\sqrt{4}} = 3/2.$$

We can find the limits using a computer algebra system. Alternatively, setting $u = e^t$, we can use the limit laws to calculate

$$\lim_{t \to \infty} \frac{e^t}{\sqrt{3 + e^{2t}}} = \lim_{u \to \infty} \frac{u}{\sqrt{3 + u^2}} = \lim_{u \to \infty} \sqrt{\frac{u^2}{3 + u^2}}$$

$$= \sqrt{\lim_{u \to \infty} \frac{u^2}{3 + u^2}} = \sqrt{\lim_{u \to \infty} \frac{1}{\frac{3}{u^2} + 1}}$$

$$= \sqrt{\frac{1}{\lim_{u \to \infty} \frac{3}{u^2} + 1}} = \sqrt{\frac{1}{0 + 1}} = 1.$$

Therefore, we have

$$\lim_{t \to \infty} P_1(t) = 1 - 1 = 0$$

$$\lim_{t \to \infty} P_2(t) = 1 + 1 = 2.$$

To predict these limits without having a formula for P, looking at the original differential equation. We see if $0 < P < 1$, then $P(P-1)(2-P) < 0$, so $P' < 0$. Thus, if $0 < P(0) < 1$, then $P'(0) < 0$, so P is initially decreasing, and tends toward the equilibrium solution $P = 0$. On the other hand, if $1 < P < 2$, then $P(P-1)(2-P) > 0$, so $P' > 0$. So, if $1 < P(0) < 2$, then $P'(0) > 0$, so P is initially increasing and tends towards the equilibrium solution $P = 2$.

CHECK YOUR UNDERSTANDING

1. False. Suppose $k = -1$. The equation $y'' - y = 0$ or $y'' = y$ has solutions $y = e^t$ and $y = e^{-t}$ and general solution $y = C_1 e^t + C_2 e^{-t}$.

5. False. This is a logistic equation with equilibrium values $P = 0$ and $P = 2$. Solution curves do not cross the line $P = 2$ and do not go from $(0, 1)$ to $(1, 3)$.

9. True. No matter what initial value you pick, the solution curve has the x-axis as an asymptote.

13. True. Rewrite the equation as $dy/dx = xy + x = x(y + 1)$. Since the equation now has the form $dy/dx = f(x)g(y)$, it can be solved by separation of variables.

17. True. Since $f'(x) = g(x)$, we have $f''(x) = g'(x)$. Since $g(x)$ is increasing, $g'(x) > 0$ for all x, so $f''(x) > 0$ for all x. Thus the graph of f is concave up for all x.

21. False. Let $g(x) = 0$ for all x and let $f(x) = 17$. Then $f'(x) = g(x)$ and $\lim_{x \to \infty} g(x) = 0$, but $\lim_{x \to \infty} f(x) = 17$.

25. True. The slope of the graph of f is $dy/dx = 2x - y$. Thus when $x = a$ and $y = b$, the slope is $2a - b$.

29. False. Since $f'(1) = 2(1) - 5 = -3$, the point $(1, 5)$ could not be a critical point of f.

33. True. We will use the hint. Let $w = g(x) - f(x)$. Then:

$$\frac{dw}{dx} = g'(x) - f'(x) = (2x - g(x)) - (2x - f(x)) = f(x) - g(x) = -w.$$

Thus $dw/dx = -w$. This equation is the equation for exponential decay and has the general solution $w = Ce^{-x}$. Thus,

$$\lim_{x \to \infty} (g(x) - f(x)) = \lim_{x \to \infty} Ce^{-x} = 0.$$

37. If we differentiate implicitly the equation for the family, we get $2x - 2y\,dy/dx = 0$. When we solve, we get the differential equation we want $dy/dx = x/y$.

APPENDIX

Solutions for Section A

1. The graph is

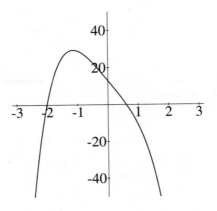

   (a) The range appears to be $y \le 30$.
   (b) The function has two zeros.

5. The largest root is at about 2.5.

9. Using a graphing calculator, we see that when x is around 0.45, the graphs intersect.

13. **(a)** Only one real zero, at about $x = -1.15$.
    (b) Three real zeros: at $x = 1$, and at about $x = 1.41$ and $x = -1.41$.

17. **(a)** Since f is continuous, there must be one zero between $\theta = 1.4$ and $\theta = 1.6$, and another between $\theta = 1.6$ and $\theta = 1.8$. These are the only clear cases. We might also want to investigate the interval $0.6 \le \theta \le 0.8$ since $f(\theta)$ takes on values close to zero on at least part of this interval. Now, $\theta = 0.7$ is in this interval, and $f(0.7) = -0.01 < 0$, so f changes sign twice between $\theta = 0.6$ and $\theta = 0.8$ and hence has two zeros on this interval (assuming f is not *really* wiggly here, which it's not). There are a total of 4 zeros.
    (b) As an example, we find the zero of f between $\theta = 0.6$ and $\theta = 0.7$. $f(0.65)$ is positive; $f(0.66)$ is negative. So this zero is contained in $[0.65, 0.66]$. The other zeros are contained in the intervals $[0.72, 0.73]$, $[1.43, 1.44]$, and $[1.7, 1.71]$.
    (c) You've found all the zeros. A picture will confirm this; see Figure A.1.

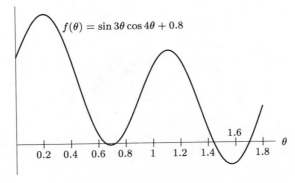

$$f(\theta) = \sin 3\theta \cos 4\theta + 0.8$$

Figure A.1

21.

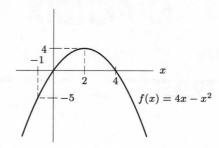

Bounded and $-5 \le f(x) \le 4$.

Solutions for Section B

1. $r = \sqrt{1^2 + 0^2} = 1, \quad \theta = 0$.

5. $r = \sqrt{(-3)^2 + (-3)^2} = 4.2$.
   $\tan \theta = (-3/-3) = 1$. Since the point is in the third quadrant, $\theta = 5\pi/4$.

9. $(1,0)$

13. $(\frac{5\sqrt{3}}{2}, -\frac{5}{2})$

17. The graph is a circle of radius 2 centered at the origin. See Figure B.2.

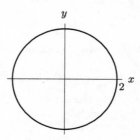

Figure B.2

21. The condition that $r \le 2$ tells us that the region is inside a circle of radius 2 centered at the origin. The second condition $0 \le \theta \le \pi/2$ tells us that the points must be in the first quadrant. Thus, the region consists of the quarter of the circle in the first quadrant, as shown in Figure B.3.

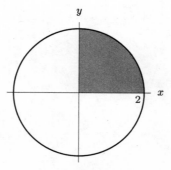

Figure B.3

25. The region consists of the portion of a circle of radius 1 centered at the origin that is between an angle of $\theta = 0$ and an angle of $\theta = \pi/4$, Therefore, the region is defined by $r \leq 1$ and $0 \leq \theta \leq \pi/4$.

29. Putting $\theta = \pi/3$ into $\tan\theta = y/x$ gives $\sqrt{3} = y/x$, or $y = \sqrt{3}x$. This is a line through the origin of slope $\sqrt{3}$. See Figure B.4.

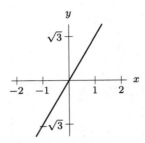

Figure B.4

33. For $r = \theta/10$, the radius r increases as the angle θ winds around the origin, so this is a spiral. See Figure B.5.

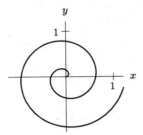

Figure B.5

Solutions for Section C

1. $2e^{i\pi/2}$

5. $0e^{i\theta}$, for any θ.

9. $-3 - 4i$

13. $\frac{1}{4} - \frac{9i}{8}$

17. $5^3(\cos\frac{3\pi}{2} + i\sin\frac{3\pi}{2}) = -125i$

21. One value of $\sqrt[3]{i}$ is $\sqrt[3]{e^{i\frac{\pi}{2}}} = (e^{i\frac{\pi}{2}})^{\frac{1}{3}} = e^{i\frac{\pi}{6}} = \cos\frac{\pi}{6} + i\sin\frac{\pi}{6} = \frac{\sqrt{3}}{2} + \frac{i}{2}$

25. One value of $(-4 + 4i)^{2/3}$ is $[\sqrt{32}e^{(i3\pi/4)}]^{(2/3)} = (\sqrt{32})^{2/3}e^{(i\pi/2)} = 2^{5/3}\cos\frac{\pi}{2} + i2^{5/3}\sin\frac{\pi}{2} = 2i\sqrt[3]{4}$

29. We have

$$i^{-1} = \frac{1}{i} = \frac{1}{i} \cdot \frac{i}{i} = -i,$$

$$i^{-2} = \frac{1}{i^2} = -1,$$

$$i^{-3} = \frac{1}{i^3} = \frac{1}{-i} \cdot \frac{i}{i} = i,$$

$$i^{-4} = \frac{1}{i^4} = 1.$$

The pattern is

$$i^n = \begin{cases} -i & n = -1, -5, -9, \cdots \\ -1 & n = -2, -6, -10, \cdots \\ i & n = -3, -7, -11, \cdots \\ 1 & n = -4, -8, -12, \cdots \end{cases}$$

Since 36 is a multiple of 4, we know $i^{-36} = 1$.
Since $41 = 4 \cdot 10 + 1$, we know $i^{-41} = -i$.

33. To confirm that $z = \dfrac{a + bi}{c + di}$, we calculate the product

$$z(c + di) = \left(\frac{ac + bd}{c^2 + d^2} = \frac{bc - ad}{c^2 + d^2} i \right) (c + di)$$

$$= \frac{ac^2 + bcd - bcd + ad^2 + (bc^2 - acd + acd + bd^2)i}{c^2 + d^2}$$

$$= \frac{a(c^2 + d^2) + b(c^2 + d^2)i}{c^2 + d^2} = a + bi.$$

37. True, since $\sqrt{a}$ is real for all $a \geq 0$.

41. True. We can write any nonzero complex number z as $re^{i\beta}$, where r and β are real numbers with $r > 0$. Since $r > 0$, we can write $r = e^c$ for some real number c. Therefore, $z = re^{i\beta} = e^c e^{i\beta} = e^{c+i\beta} = e^w$ where $w = c + i\beta$ is a complex number.

45. Using Euler's formula, we have:

$$e^{i(2\theta)} = \cos 2\theta + i \sin 2\theta$$

On the other hand,

$$e^{i(2\theta)} = \left(e^{i\theta} \right)^2 = (\cos \theta + i \sin \theta)^2 = (\cos^2 \theta - \sin^2 \theta) + i(2 \cos \theta \sin \theta)$$

Equating real parts, we find

$$\cos 2\theta = \cos^2 \theta - \sin^2 \theta.$$

49. Replacing θ by $(x + y)$ in the formula for $\sin \theta$:

$$\sin(x + y) = \frac{1}{2i} \left(e^{i(x+y)} - e^{-i(x+y)} \right) = \frac{1}{2i} \left(e^{ix} e^{iy} - e^{-ix} e^{-iy} \right)$$

$$= \frac{1}{2i} \left((\cos x + i \sin x)(\cos y + i \sin y) - (\cos(-x) + i \sin(-x))(\cos(-y) + i \sin(-y)) \right)$$

$$= \frac{1}{2i} \left((\cos x + i \sin x)(\cos y + i \sin y) - (\cos x - i \sin x)(\cos y - i \sin y) \right)$$

$$= \sin x \cos y + \cos x \sin y.$$

Solutions for Section D

1. (a) $f'(x) = 3x^2 + 6x + 3 = 3(x + 1)^2$. Thus $f'(x) > 0$ everywhere except at $x = -1$, so it is increasing everywhere except perhaps at $x = -1$. The function is in fact increasing at $x = -1$ since $f(x) > f(-1)$ for $x > -1$, and $f(x) < f(-1)$ for $x < -1$.

(b) The original equation can have at most one root, since it can only pass through the x-axis once if it never decreases. It must have one root, since $f(0) = -6$ and $f(1) = 1$.

(c) The root is in the interval $[0, 1]$, since $f(0) < 0 < f(1)$.

(d) Let $x_0 = 1$.

$$x_0 = 1$$
$$x_1 = 1 - \frac{f(1)}{f'(1)} = 1 - \frac{1}{12} = \frac{11}{12} \approx 0.917$$
$$x_2 = \frac{11}{12} - \frac{f\left(\frac{11}{12}\right)}{f'\left(\frac{11}{12}\right)} \approx 0.913$$
$$x_3 = 0.913 - \frac{f(0.913)}{f'(0.913)} \approx 0.913.$$

Since the digits repeat, they should be accurate. Thus $x \approx 0.913$.

5. Let $f(x) = \sin x - 1 + x$; we want to find all zeros of f, because $f(x) = 0$ implies $\sin x = 1 - x$. Graphing $\sin x$ and $1 - x$ in Figure D.6, we see that $f(x)$ has one solution at $x \approx \frac{1}{2}$.

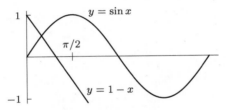

Figure D.6

Letting $x_0 = 0.5$, and using Newton's method, we have $f'(x) = \cos x + 1$, so that

$$x_1 = 0.5 - \frac{\sin(0.5) - 1 + 0.5}{\cos(0.5) + 1} \approx 0.511,$$

$$x_2 = 0.511 - \frac{\sin(0.511) - 1 + 0.511}{\cos(0.511) + 1} \approx 0.511.$$

Thus $\sin x = 1 - x$ has one solution at $x \approx 0.511$.

9. Let $f(x) = \ln x - \frac{1}{x}$, so $f'(x) = \frac{1}{x} + \frac{1}{x^2}$.
Now use Newton's method with an initial guess of $x_0 = 2$.

$$x_1 = 2 - \frac{\ln 2 - \frac{1}{2}}{\frac{1}{2} + \frac{1}{4}} \approx 1.7425,$$
$$x_2 \approx 1.763,$$
$$x_3 \approx 1.763.$$

Thus $x \approx 1.763$ is a solution. Since $f'(x) > 0$ for positive x, f is increasing: it must be the only solution.